THEY ALL LAUGHED...

THEY ALL LAUGHED...

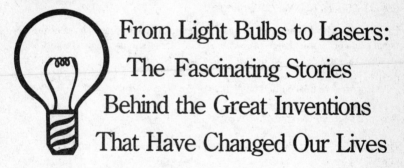

From Light Bulbs to Lasers:
The Fascinating Stories
Behind the Great Inventions
That Have Changed Our Lives

IRA FLATOW

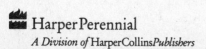

HarperPerennial
A Division of HarperCollinsPublishers

Acknowledgments for photographs and illustrations follow the index.

A hardcover edition of this book was published in 1992 by HarperCollins Publishers.

HarperCollins books may be purchased for educational, business, or sales promotional use. For information please write: Special Markets Department, HarperCollins Publishers, Inc., 10 East 53rd Street, New York, NY 10022.

First HarperPerennial edition published 1993.

Designed by Irving Perkins Associates

The Library of Congress has catalogued the hardcover edition as follows:
Flatow, Ira.
   They all laughed — : from lightbulbs to lasers, the fascinating stories behind the great inventions that have changed our lives / Ira Flatow — 1st ed.
       p.   cm.
   Includes index.
   ISBN 0-06-016445-X (cloth)
   1. Inventions—History.   I. Title.
T212.F53   1992
609 — dc20   91-58336

ISBN 0-06-092415-2 (pbk.)

   05 04 03   PS/RRD 20 19 18 17

To Sam, Anna, and Abigail

Contents

Acknowledgments

In the course of writing this book, I was fortunate enough to be welcomed into the close-knit community of history-of-science writers. These veteran historians delighted in sharing obscure facts and secrets with a fledgling history writer. They were my source of joy and inspiration. When my resourcefulness ran low, they refreshed my spirit with unselfish direction and renewed my interest with a teasing "Did you hear about . . ." tale of discovery and invention.

My utmost gratitude goes to Richard Q. Hofacker, Jr., Special Projects Writer, AT&T Bell Laboratories, Short Hills, N.J., for his tireless and resourceful help; to George Wise, Specialist-Communications, General Electric Company, Schenectady, N.Y., and Eliot Sivowitch, Museum Specialist, Smithsonian Institution, Washington, D.C., who helped me wade through the history of electricity; to Dr. Theodore Bernstein, Consulting Engineer, Madison, Wisc., for electrical expertise and Dr. Terry S. Reynolds, Head, Department of Social Sciences, Michigan Technological University, Houghton, Mich., for help in sharing hard-to-get-photographs; to Glenn Zorpette, Senior Associate Editor, *IEEE Spectrum*, New York, N.Y., for steering me to the gun director and the Colossus computer.

Dr. Earl Warrick, retired Dow Corning scientist and creator of the original Silly Putty formula, was kind enough to fill me in on the gaps in Silly Putty history and to provide original photographs. Rose Beaudry of Waring Products Division, New Hartford, Conn., and Pete Kiefer, Curator of the Fred Waring Archives at University Park in Pennsylvania, helped me make sure I spelled Blendor correctly and secured valuable photos. Also thanks to Robert W. Grupp, Manager of External Communications, Dow Corning Corporation, Midland, Mich.; to Richard G.

Kuhl, VELCRO® USA, Manchester, N.H.; to Jon Caroulis, University of Pennsylvania, Philadelphia, Office of News and Public Affairs, for the photo of ENIAC; to the Computer Museum, Boston, for the photo of Colossus.

The *American Heritage of Invention & Technology* is a magazine of priceless historical information. Thank you, Fred Allen, Editor, for your encouragement and expertise. Thanks also to Julie Maddocks of the David Sarnoff Research Center, Princeton, N.J.; to Ellen Denuto of the Paterson Museum, N.J.; to Diane Currie of Du Pont, Wilmington, Del.; to Ann Humbert of Amana Refrigeration, Inc., Amana, Ohio.

When I needed material about George Eastman, Alison Hofland of Creative Communications Services, Encinitas, Cal., and Lois Gauch of Eastman Kodak, Rochester, N.Y., came to my rescue. When I needed information about Westinghouse, historian Charles Ruch of Pittsburgh flooded me with good stuff.

Without the spark supplied by Jean-Marie Asselin of Mississaugua, Ontario, I would have never looked into the fascinating history of the fax machine. Thanks to Dan Keller of Wyndmoor, Pennsylvania, for the fax photos from Milan and to Orazio Curti, Il Direttore Generale F.F., Museo Nazionale della Scienza e della Tecnica, Milan, Italy, for allowing me to pry open his exhibit case and get to the ancient fax machines. Grace Sun of the Asia Society in New York City was invaluable in translating the Chinese pantelegraph cover sheet on short notice.

My eternal gratitude goes to Dorthea Nelhybel, Librarian, Burndy Library, Norwalk, Conn. Following the death of its magnificent founder and friend, Bern Dibner, the library is closing up shop and moving. The Burndy Library was one of the truly outstanding science and technology resources. It will be terribly missed. Thank you, Dorthea, for allowing me to call the library home.

Thanks to Carlos Toro, who has gone beyond the call of duty in assisting me with this book; to my literary agent, Carol Mann; and to my editor at HarperCollins, Cynthia Barrett, a delight to work with, who never hung up the phone on me despite my interminable ramblings.

Foreword

The idea for this book was hatched at the Franklin Institute in Philadelphia. Schmoozing with Charles Penniman of the Institute, I mentioned I was looking for grist for a second book based on the history of science and technology, one that would look at how ideas and inventions developed but with an unusual twist: I was looking for uncommon knowledge, stories about common inventions and discoveries that had unusual beginnings.

Without missing a beat, Penniman said, "Franklin. Look into the story of Ben Franklin and his kite. The 'experts' all laughed when he proposed the idea that lightning was in fact electricity. And even more interestingly, he wasn't the first to perform the famous experiment that proved it, you know. It was done in Europe first."

"Really? I didn't know that!"

"Sure," he replied. "We even mention it in our lecture we give about Franklin to tourists."

Now this was news. Ben Franklin is a Philadelphia icon. One would think the city would shun any adverse publicity for its patron "saint." Any tidbit of gossip that dulls the luster on the Franklin shine would be a closely guarded secret to an institute that bears his name. Unless of course it's true. For Penniman to offer this morsel of a lead must have meant he thought it was a great yarn.

And it was. Turns out that Franklin was frustrated in his efforts to draw electricity from clouds, so he described to other scientists—mainly in Europe—his unproven method of drawing electric current via a long metal rod poked into the sky. The French tried it first and succeeded, proving Franklin's theory of electricity. But this French achievement is not what the world remembers. We don't learn about the French lightning experiment in school. What the world remembers is Franklin's kite-flying experiment—Franklin's attempt at duplicating the French feat but with flair: via a

metal-tipped kite, not a rod (an early public relations triumph?).
Franklin's tale set me thinking about other bits of historical common
knowledge we take for granted. History is filled with myths and
legends of who did what first, and how. Was Edison's light bulb
really invented the way we think it was? How about the telephone?
Computer games?

The ideas for topics began to jell. Mining the archives for other
unusual stories yielded unexpected historical nuggets in unusual
places. A letter to the editor in *PC Magazine* quietly pointed out
that the fax machine was invented *before the telephone* by an Italian
priest. Could this really be true? The fax older than the phone? I
had to see one for myself. But where to find one? The Smithsonian had one locked up and lost in storage in some immense
warehouse (if you're familiar with the last scene of *Raiders of the
Lost Ark*, you know what I'm talking about). Only an act of
Congress, literally, could pry it loose. Then, purely by accident, I
got lucky. While on vacation in Milan, Italy, I played a hunch and
found the *pantelegrafo* proudly on display at the National Museum
of Science and Technology.

The more I looked into the history of inventions, the more I
learned how *crazy* the inventors were considered. People laughed
at Bell and his telephone. The navy couldn't fathom a use for the
submarine. And the inventors had epic fights over the right to be
called "first" with an idea. The telephone, light bulb, radio, television, and computer were all "invented" by lots of people who added
bits and pieces to a large puzzle.

This book is an attempt to demythologize the world of invention
and discovery, to prove that truth *is* stranger than fiction and more
amazing than any hype flowing from the typewriters of Hollywood
mythologizers. You'll see why Americans idolize Edison but some
British think him an also-ran. You'll understand how a wasp can
change history and how a dream helped turn the tide of war.

Each story is a tale about people. Yes, the table of contents lists a
host of inventions and discoveries. But at the heart of each one of
these objects is the story of the people behind the inventions and
ideas, people who have certainly earned the title of inventor but

who above all are mostly inquisitive souls who share one thing in common: They refuse to take no for an answer.

What I learned from my research into invention is that it takes no particular schooling to be good at invention. There is no place to go to learn invention. Invention, I found, comes to people who are willing to hear the word no everyday from potential backers and who can rebound from the continual snickering behind their backs when things go wrong. To be a successful inventor, one needs a very thick skin.

We've all experienced the feeling. Any person who has banged the side of a TV set to get it working or kicked the refrigerator to stop its rattling knows the feeling when you've come to the end of the rope. When conventional methods just won't do the job. Well, inventors are like that. Some are deep thinkers, trained in the rigid discipline of scientific analysis. And in most cases, they begin their work armed with a good idea of how things should turn out. But they get to the point where no matter what they do, the damned thing just isn't working. The experiment is not paying off. The hours, days, months, years in the laboratory are just not paying dividends.

Then that primordial instinct common to all humans surfaces: the urge to strike back. Hit the glassware. Punch out the power supply. But inevitably the successful ones regain their senses and have a whack at it one more time. Inventors don't always produce the objects they set out to make. Bell didn't set out to invent the telephone, nor Edison the phonograph. Each was looking for a device that would make it easier for businessmen to transmit and store facts and figures.

Some inventors are interested in following diverging paths when they find them. Others work with blinders on and follow a focused path. Bell and Edison are good contrasts. When opportunity knocked, Bell had the flexibility to change directions and follow the path to the phone, willing to give up his search for a better telegraph. Edison, rigid in his methodology, stumbled onto a phenomenon that later served as the foundation for the vacuum-tube revolution. Instead of diligently following that new lead, he gave it a cursory glance and went back to tinkering with his bulbs.

I learned that inventors have a childlike curiosity; they're not afraid to ask silly questions. I call this the "quack like a duck" method of discovery. When something looks like a duck and walks like a duck, but doesn't quack like a duck, they're not afraid of crying foul. When Constantine Fahlberg noticed that a slice of bread he was eating at dinner tasted unusually sweet, the easy solution would have been to merely assume that sugar had accidentally fallen into the dough or that he'd mistakenly used the sugar instead of the salt shaker. But that answer didn't quack. Instead Fahlberg went back to his chemical laboratory and tasted everything in sight until he found the sweet chemical powder he'd gotten on his hands—later called saccharin.

Similarly, when Percy Spencer noticed that a candy bar had melted in the pocket of the lab coat he'd worn all day, he could have assumed, as others did, that the chocolate was softened from body heat. But that answer didn't make duck noises either. Spencer went on to describe how accidental radiation coming from his laboratory device could be turned into a microwave oven.

The problem every writer has is knowing when to stop, how deep to dig for *the* real inventor. Giving credit to any one person as being the father (or mother) of an innovation is almost impossible. In many cases, ideas are constantly being refined over the years. The "father" of television in 1928 may become the inconsequential inventor of 1958. Sometimes popular culture—newspapers, films, television—may have conferred the coveted moniker upon some charismatic but undeserving publicity seeker. Other times the credit can remain locked in controversy for many years, until the courts make someone an instant zillionaire. The patent for the computer-on-a-chip was awarded in 1990 to an "unknown" engineer, Gilbert P. Hyatt, who battled the patent office for twenty years. Gil Who? is now starting to collect tens of millions of dollars in royalties.

The problem was best expressed by Eliot Sivowitch, Museum Specialist in the Division of Electricity at the National Museum of

American History (Smithsonian Institution) in Washington, D.C. After a long, exasperating conversation over who deserves credit for inventing the laser, Eliot told me of his Sivowitch Law of Firsts: "Whenever you prove who was first, the harder you look you will find someone else who was more first. And if you persist in your efforts you find that the person whom you thought was first was third. Someone will appear on the scene who was more first than you thought was first in the first place."

Ira Flatow
Stamford, Conn.

1

....

Franklin, the Modern Prometheus

The kite is to be raised when a thunder-gust appears to be coming on (which is very frequent in this country) . . .

> —BENJAMIN FRANKLIN, describing
> how to fly a kite in a storm,
> October 1, 1752

ONE OF THE greatest myths of American history is the one told about Ben Franklin and his kite. We all grew up learning that good old Ben went out one stormy day and flew his kite in a thunderstorm, and when the kite was hit by lightning, Franklin discovered electricity.

Fortunately, Ben was a lot luckier than that. If lightning had actually struck his kite, Ben might not have been around to sign the Declaration of Independence and write *Poor Richard's Almanac*— and the rest of the world would have been much poorer for his stupidity.

Oddly enough, Ben wasn't the first person to carry out his lightning experiment. And while Franklin carefully described the steps to building and flying a kite in a storm in a letter dated October 1, 1752, no personal diary of Franklin describing his own kite-flying experience exists. He never wrote up the experiment as a good scientist—which he was—normally would. Rather, all we have to go on is his oral description written down by an acclaimed scientist of the day, Joseph Priestley, which for most reputable historians is good enough.

Perhaps for that reason the tale has grown into mythic proportions. If so, it goes along with the almost God-like reputation Franklin amassed during his tenure as America's first and premier electrical genius.

THE SENTRY-BOX

A visit to Boston in 1746 brought Franklin to a demonstration of electricity by a Scottish lecturer, Dr. Adam Spencer. Franklin became so enamored with the sparks and cracks of the high-voltage world of static electricity that he immediately decided he had to own Spencer's equipment. Making him an offer he couldn't refuse, Franklin bought all of his paraphernalia. Peter Collinson, a friend in England, sent him some more, along with notes about the electrical experiments being conducted there.

The feisty colonist took issue with the thinking of Europe's leading scientists of the day (called natural philosophers) who claimed that electricity was really of two types. They claimed that static electricity produced by rubbing a glass rod would attract a pith ball. But electricity produced by rubbing a resinous rod would repel the ball because each kind of rod produced a different form of electricity.

Hogwash, thought Franklin. It's all the same stuff. One rod simply has more of it or less of it. Electricity flows, he said, from greater charge to lesser charge. You can pile it up here and let it snap back there when you draw the spark of an electric fire. The spark is the way electricity evens up all those charges that have been separated.

Franklin's experiments set Europe buzzing. Who was this young upstart who'd spent only five years of his life studying electricity? Nevertheless, this colonial printer did make sense, and his experiments couldn't be refuted. More importantly, they revolutionized the world of electricity and made Franklin's reputation as a world-class authority on the subject.

THE IDEA BEHIND THE KITE

Having already made a fortune in the printing business at the age of forty-two, Franklin turned his attention to experiments in electricity. He wanted to know if lightning and static electricity were made of the same stuff. Experiments were performed in Philadelphia in 1749 to test his idea that lightning was an electrical discharge from cloud to cloud and from cloud to earth. But he still wanted to find some way of capturing the electrical "fire" from storm clouds and bottling it up so it could be studied alongside common, earthly electrical charges. So he proposed an experiment for drawing electrical charges from clouds:

> On the top of some high tower or steeple, place a kind of sentry-box big enough to contain a man and an electrical stand. From the middle of the stand let an iron rod rise and pass bending out of the door, and then upright 20 or 30 feet, pointed very sharp at the end. If the electrical stand be kept clean and dry, a man standing on it when such clouds are passing low, might be electrified and afford sparks, the rod drawing fire to him from a cloud.

Lest anyone be frightened off by the possibility of being fried by a lightning bolt, Franklin offered these consoling words:

> If any danger to the man should be apprehended (though I think there would be none) let him stand on the floor of his box, and now and then bring near to the rod the loop of wire that has one end fastened to the leads, he holding it by a wax handle; so the sparks, if the rod is electrified, will strike from the rod to the wire and not affect him.[1]

Franklin described a good method of insulating, via wax, the subject from the wire. He also had the experimenter step down from the insulated box. So Franklin knew of the dangers of an errant lightning bolt and warned others to protect themselves.

The sentry-box experiment was published in London in 1751.

The Abbé Nollet, the most prominent French electrical experimenter of the time, turned his nose up at the experiment. He didn't believe it could be done. Fortunately, the French king, Louis XV, loved the idea and encouraged his scientists to perform the experiment and confirm Franklin's notions about the similarity of lightning and electricity. A garden in Marly, eighteen miles from Paris, was chosen for the site. On May 10, 1752, Jean François D'Alibard erected a pointed metal rod forty feet high, resting on an insulated table. He waited for a storm cloud and was not disappointed. One conveniently appeared and passed over. According to Franklin's own accounts, at least one person cautiously drew near and attracted sparks from it using an insulated wire loop, just as Franklin had described.

The successful experiment was repeated time and time again throughout the summer of 1752 by wildly enthusiastic French scientists. Attempting to outdo one another, they gradually increased the height of the bar to ninety feet. Not to be outdone by their rivals across the Channel, British experimenters tried—and failed—to attract their own sparks. A damp summer of 1752 was blamed. However, the British enthusiastically welcomed the French news. "Abundance of persons of indisputable credit were eyewitness of the effects it [the rod] produced," proclaimed a report in the *London Magazine*, "from whence it is now demonstrable that the effects of lightning and electricity are the same."[2]

INSTANT FAME FOR FRANKLIN

The results of the experiments made Franklin an international hero. That electricity could be drawn from a cloud, stored in large glass and metal vessels (called Leyden jars), and appear in experiments to behave exactly as electricity did on terra firma proved Franklin a genius. They loved him in Europe. Praises ranged from the Arno to the Seine. The fledgling American colonies had truly produced a world-class scientist. "All our European electricians must doff their hats to this American,"[3] wrote the Italian Francesco Algarotti.

"A modern day Prometheus" is how German philosopher Immanuel Kant described him, alluding to the mythical Prometheus who drew fire from the heavens. Back at home, Franklin's friends were excited as well. John Adams said that Franklin's reputation was "more universal than that of Leibniz or Newton, Frederick or Voltaire; and his character more esteemed and beloved than all of them."[4]

And all this without even having flown a kite!

THE FAMOUS KITE EXPERIMENT

For his own part, the originator of the idea had yet to get into the act. Having not heard of the success of the electricians in Europe that summer of '52, Franklin was eagerly awaiting the opportunity to try out his own experiment using a metal pole atop the steeple of the tall Christ Church in Philadelphia. He believed the rod had to be elevated to great heights to draw off electrical charges. But construction of the steeple had been going slowly, and Franklin lost patience. Not realizing that just a moderate height would do the trick, Franklin struck on the idea of using a kite to carry the rod aloft.

In June 1752 Franklin's patience ran out. A month after the experiments in Europe had already confirmed his theory, Franklin embarked on his kite experiment. Fastening a cross out of two sticks and stretching a large silk handkerchief across them, Franklin created his legendary kite. Affixed to the upper tip was a sharp metal wire. At the other end of the string were a metal key and an insulating silk ribbon by which Franklin held on to the kite string. Waiting for a thunderstorm to approach, the experimenter, accompanied only by his son, walked out into a field and to a shed that afforded shelter from potential rain. Once the kite was airborne a considerable time elapsed before a thunder cloud approached. According to Priestley's account:

One very promising cloud had passed over it without any effect; when, at length, just as he was beginning to despair of this contrivance of his, he observed some loose threads of the hempen string to stand erect and to avoid one another, just as if they had been suspended on a common conductor. Struck with this promising appearance, he immediately presented his knuckle to the key, and (let the reader judge of the exquisite pleasure he must have felt at the moment) the discovery was complete. He perceived a very evident electric spark.[5]

Franklin continued to draw sparks from the key and collected electricity in Leyden jars. What must be emphasized is that at no point was Franklin's kite *hit* by lightning. No bolt came out of the cloud and struck the kite. This was not Franklin's intention. Franklin wanted to draw charges from the cloud down the string. The string remained charged until Franklin, grounding himself, brought his knuckle to the key and drained the charges off into the ground via his body.

Historians are a bit puzzled by the secrecy surrounding the experiment. Why did he inform only his son of his intentions? Was Franklin fearful of ridicule that might follow an unsuccessful attempt? Was he fearful of foolishly electrocuting himself in a public display?

Perhaps Franklin was just lucky he didn't kill himself. At least one other was not. Trying to repeat the sentry-box experiment, Russian physicist G. W. Richman was killed in St. Petersburg in 1753 when "a palish blue ball of fire, as big as a fist, came out of the rod" and struck him in the head. Richman died instantly from the lightning bolt and became the first martyr to the new age of electricity.

THE BIRTH OF THE LIGHTNING ROD

Franklin's own description of the event was not full of important details of time, place, or procedures that usually accompany historic experiments. In 1767 Joseph Priestley published his own detailed account of Franklin's kite experiment of June 1752, hinting

Franklin's Kite-flying Experiment. Franklin designed his kite experiment when he grew impatient waiting for a church steeple to be erected. Hoping to draw off electricity from a cloud via an iron rod affixed to the steeple, Franklin settled upon using a kite to carry the rod instead. Franklin can be seen drawing off electric charge into an "electric bottle" (Leyden jar) pointed at the key. Notice that lightning *does not* strike the kite!

that he had gotten the particulars from Franklin himself. Franklin published just a description of the experiment, not an account of it, in the October 19, 1752, issue of the *Pennsylvania Gazette*. In that same issue, he promoted an upcoming article in *Poor Richard's Almanac* for 1753 entitled "How to Secure Houses, etc. from Lightning." Franklin was a firm believer in lightning rods. He began recommending them in 1751. And these new experiments in lightning bore out his belief that lightning rods could effectively protect the public from bolts from the sky.

The purpose of the rod, he wrote, was not to be struck by lightning but to prevent strikes from happening. The rod was supposed to continually draw off charges from the clouds and shunt them to the ground before they became strong enough to form a lightning bolt. As Franklin put it: "Draw the electrical fire silently out of the cloud before it came nigh enough to strike."

Despite Franklin's reputation, Europeans did not rush to erect lightning rods, either because they didn't believe in them or because they didn't understand how they worked. Those who had witnessed the Marly experiment were fearful that any lightning rod attached to their homes would "attract" lightning that would otherwise not be there.

The master himself lost no time installing a rod in his own home, in 1752. Affixed to the top of the chimney and running nine feet above, Franklin's iron rod threaded its way through the house in a less than benign fashion. Incessantly the tinkerer, Franklin rigged the wire to pass through a home-made bell, the construction of which would hardly pass today's electrical building codes. A six-inch gap was cut into the wire and a bell attached to each end. A brass ball was hung between the bells. Electrically charged clouds passing overhead made the ball dance between the bells, setting off a ringing that must have sounded like music to Franklin's ears.

Never sufficiently fearful of the bolts, Franklin described his delight in being "one night awaked by large cracks on the staircase." The bell, instead of ringing, was being repelled as "fire passed . . . sometimes in a continued, dense, white stream seemingly as large as my finger, whereby the whole staircase was inlightened as with sunshine so that one might see to pick up a pin."

In a short time lightning rods were sprouting on buildings all over the continent. Franklin had proven his point. Soon newly independent colonists would be erecting the rods on rooftops.

THE KITE AND THE AMERICAN REVOLUTION

In the long run, the sentry-box and kite experiments may have been much more important to the history of the United States than to the annals of electricity. Certainly, out of the experiments came a very useful invention—the lightning rod—that saved the lives of countless people. Perhaps more importantly, Franklin's enormous popularity in Europe and especially in France, where the sentry-box experiment was first done, may have helped America win its independence.

Franklin wrote all of his ideas in letters to his friend in London, Peter Collinson, who reported Franklin's findings to the Royal Society. Collinson collected all these letters and published them in 1751 in a book that was widely read and caught the eye of the last king of France. Had it not been for King Louis's fondness for Franklin, he might have turned a deaf ear to the pleas of colonists for French help in fighting the British. Support of a bloody revolution against a monarchy was not the king's style; think of the ideas it might spark in his own French subjects. And a plea from a common Philadelphia printer would hardly have reached the royal ears.

But Franklin's personal reputation and closeness to the king would prove to be decisive in bringing His Majesty to side with the revolutionaries. And Americans might well adopt a tradition of flying a kite each Fourth of July to celebrate their independence that owes a measure of its success to an experiment in electricity.

2

....

Whose Light Bulb? Edison in a New Light

*Fifteen years ago, I used charred paper and card in the
construction of an electric lamp on the incandescent principle.
I used it, too, in the shape of a horse-shoe, precisely as you say
Mr. Edison is now using it.*

> —JOSEPH SWAN, British inventor
> and rival of Edison, January 1,
> 1880

WHAT DO HIRAM Maxim, Joseph Swan, Thomas Edison, and more
than a dozen other inventors have in common? They all invented
electric light bulbs. But unless you're British and insist that Joseph
Swan deserves credit for the invention, you've probably never
heard of any name associated with the bulb besides Edison's.
Chances are, if you examine a list of inventors who have toiled to
capture electric light in a glowing glass bulb, you'll be shocked (no
pun intended) to see just how much work was going on in other
laboratories outside Menlo Park in New Jersey (see "The Light
Bulb Hall of Fame," p. 26)

Edison was certainly not the first to come up with the incandescent light bulb idea. And contrary to popular opinion, the key
ingredient he used for his light bulb—carbon—was certainly not
unique. Carbon had been an ingredient of experimental light bulbs
fifty years before Edison.

Why is it then that the world accepts Menlo Park as the bulb's birthplace and Edison as *the* sole light bulb inventor? It's true that Edison received numerous patents for his light bulbs. But so did others. The mid-nineteenth century was an inventor's paradise.* (As readers will note throughout this book, having a patent doesn't hold much water in the political race to be named as *the* inventor.) To historians who recognize Edison as the inventor the answer involves the difference between having just an idea—an invention—and having a way of implementing the idea. An invention is worthless if it only functions under laboratory conditions.

At least three or four serious inventors, in England, France, and the United States, were working on the incandescent lamp in the 1870s. They had the right ingredients and had functioning light bulbs. Joseph Swan had lit residences with his British bulb. Hiram Maxim, a serious American inventor and intense rival of Edison's, had filed for incandescent light patents in 1878 and 1879 and had carbon incandescent lamps burning for twenty-four hours at a time. But while Swan, Maxim, and other successful inventors succeeded in producing workable bulbs, theirs functioned only on a small scale, where but a few light bulbs were needed for illumination.

Only Edison designed his lamp, from the beginning, to be part of a total electrical system the size of a city, complete with electric dynamos to produce the electricity and wires and fuses to distribute and control it. Only Edison discerned that the lamp and the system had to work as a unit and had to match. And only when Edison realized that his bulb would have to work outside the lab and be part of a total system did it dawn on him how to create a properly working bulb.

In addition, it was Edison's enormous wealth, influence, and power that allowed him to create the entire system from scratch in

---

* Colonial period mapmakers, surveyors, mathematicians, and seamen busily designed their own amateur instruments to explore and measure their immense country. This trend continued into the middle of the nineteenth century. The Northeast became a mecca for electrically minded inventors. For more, see Silvio Bedini's excellent book *Thinkers and Tinkers*, published in 1975 by Scribner's.

A Glowing Money Maker.
Sketched as a glowing money mogul, Edison's new light bulb of 1888 was perceived by Edison associate Francis Upton as a cash cow for the company and its investors.

his New Jersey laboratories, set up a power station to light New York City with his new bulbs, and influence an eager press and public into believing his bulb to be superior to all others.

THE WIZARD OF MENLO PARK

By the time Edison set his mind on the light bulb, he had already become very famous. Sitting and watching a stock ticker spit out Wall Street quotes, he commented how easily an improved version could be made. And he made one in 1869. Shut out of the invention of the telephone—and any royalties from it—by Bell's patent, Edison did the next best thing: He designed an improved carbon transmitter (mouthpiece) that became the mainstay of the phone system. Intrigued with finding a method of recording telegraph messages, Edison invented a means of recording speech. This last invention, the phonograph, brought him instant fame.

Whatever Edison said he would do, he did. The track record of this brash youngster from America's heartland was nothing short of spectacular. From his laboratory in suburban New Jersey, he promised a "minor invention" every few days and a major one every six months or so. He had the phonograph, the stock ticker, and the carbon telephone transmitter to back up his bravado. He also had the heavy financial backing of Wall Street financiers who recognized a "technology factory" when they saw one.[1]

So when the Wizard set his sights on a project, few people doubted he could pull it off. And the next project brewing in Edison's mind that summer of 1878 was finding a way of bringing electric lighting into everyone's home. After observing the workings of an electric dynamo Edison returned home, digested what he had seen, and came to the conclusion that he could install underground wires and light up all of lower Manhattan using yet to be invented incandescent light bulbs. This idea he called "a big bonanza." To the supremely confident—some say cocky—inventor, the light bulb would be a piece of cake. Just a matter of a few weeks, he said. In truth, Edison would sweat for almost two years.

CARBON ARC LIGHTING

Understand one thing at this point. Electric lighting was not a foreign concept; electric lights were common sights. Two well-known ways of making light by electricity existed. In the first, an electric arc was created by bringing two carbon rods close to each other and sending loads of electric current through them. A giant, blindingly white spark would jump across the gap creating an enormous amount of light, on the order of four thousand candlepower.

To get some idea of how bright electric arc lighting is, think of those searchlights you see lighting up the night skies outside movie theaters (they used to be employed to illuminate enemy war planes flying overhead). That intense beam is made by electric arc lighting. A less intense electric arc—but equally blinding—is produced by welders in their electric arc sets. Notice the kind of eye protection they must wear.

In the 1870s electrical generators were strong enough to produce electric arc lighting. They powered lighthouses and lit up public assembly areas on the streets. But you could hardly bring one into the house. It would blind everyone. Gaslight, the mainstay of illumination at that time, produced on the order of about ten to twenty candlepower. Compare gaslight to the four thousand candlepower of arc lights, and the problem becomes obvious.

The other method of electrical lighting is called incandescent, or light produced by heat. If you take a substance and pass enough electricity through it, it will become hot enough to give off light. That's great. But in the process you run the danger of melting or burning up the substance. Every inventor who had tried to make an incandescent lamp—with inventions dating back to 1823—wound up with a melted puddle of metal or a material that caught on fire. For example, platinum, which had a high melting point, did not catch fire but got too hot and melted. Carbon, which did not melt, continually caught fire even in a partial vacuum.

EARLY LIGHT BULB EFFORTS

For thirty years names like Starr, Petrie, de Changy, Farmer, Koslof, and Roberts, from countries like France, England, Russia, Belgium, and the United States, would come and go as they all attempted to harness the electric light.

In 1841 an English inventor, Frederick De Moleyns, patented an incandescent bulb using both platinum and carbon. American J. W. Starr patented two lamps, one using carbon and the other platinum, in 1845. He showed them around England to a very receptive audience, but both proved to be worthless. One interested party, Joseph Swan, an English chemist, picked up where Starr left off. He enclosed paper in a vacuum and produced a crude incandescent light. But the carbon crumbled and Swan retired from the light-bulb-inventing business in 1860 (only to resurface later when better vacuums came to his attention).

The problem had inventors and analysts literally scratching their heads for new ideas. Could it actually be possible to "subdivide" the intense electric arc lighting and bring it into the home or to produce an incandescent lamp that burned long enough to be useful? Journalists at the time thought not. At least one newspaper dubbed the solution "unknown to science."

Onto this stage rode the man on the white horse, Thomas Edison. Here was a challenge tough enough to fit Edison's press clippings. Fixing himself in Menlo Park, the boy genius—barely in his thirties—set up a whirlwind few days of experimentation. Hiring an astute English assistant, Charles Batchelor, Edison dug into his platinum supply and fashioned burners* out of platinum wire. Sealed in an evacuated glass bulb, these spiral burners had an ingenious switch built into them. As temperatures rose in the platinum, the switch would open, cutting off electricity and lowering the temperature.

EDISON BRAVADO

To Edison, a veteran of the world of telegraphy, victory in the lamp war was simply a matter of finding the right kind of switch to regulate the temperature of the burner. As an expert in switches (after all, what are telegraphs made of?), Edison boasted he'd have the problem licked in just a few weeks. Without even an operating light bulb to exhibit, Edison bragged to the newspapers on October 20, 1878, "I have just solved the problem of the subdivision of the electric light."

The inventor's supreme confidence in his own brilliance was enough to control the stock market. Talking about the death of gas lighting, speculators drove down the price of gas stocks. Backers were falling over themselves to jump on the bandwagon. J. P. Morgan and the Vanderbilts opened their coffers, and out flowed the cash used to found the Edison Electric Light Company. The

---

* It is tempting to call these burners filaments for the sake of convenience. But the word filament would eventually be defined by the courts as meaning long, extremely thin threads of material, usually made out of carbon, which these were not.

new company financed the expansion of Menlo Park and assured any and all funding needed to develop the new lamp.

With renewed life, Menlo Park kicked into high gear. Technicians busily modeled new lamps. Engineers fashioned new electric generators. Dozens of new regulator switches were attached to platinum burners, each designed to cut off the power should the temperatures go too high. The burners themselves were remodeled and reshaped; the aim was to find the best shape that provided the most light while producing the least heat.

THE BULBS WON'T WORK

Work was going like gangbusters at the lab. Things couldn't be busier. There was just one detail that couldn't be overcome—the lamps didn't work. Nothing the technologists did could solve the problems built into the lamps. Either they flickered too much from the rapid on-off switching of the heat-sensitive regulators, or the platinum melted. If, responding to the heat problem, the current was turned down, the light became too faint. The design was just too laden with bugs. It was hard keeping such bad news from upsetting the investors. By November 1878, when word of continuing failures was out, Edison was forced to admit that he needed a new line of attack. And some new brain power. He hired a young physicist from nearby Princeton, Francis Upton. Upton didn't bring a new magical bag of scientific tricks. But he brought the scientific discipline that the self-educated Edison lacked. Edison was a "try, try again" inventor; he would blindly try a new burner, test it, throw it away when it didn't work, and try a new one: a very unselective, inelegant, "brute force" process. Upton's more scientific mind led him to learn from the research of his rivals. Why waste time repeating their mistakes?

A CHANGE IN TACTICS

Immediately Edison's staff began searching through old patents, gaslight journals, and the works of their competitors. Whereas

Edison had previously been preoccupied with building one prototype after another until he found one that worked, he now realized how little he understood about the materials he was working with. And to his credit, he stopped the inventing process and had Upton and his associates concentrate on learning everything they could about the lamp materials. They tested different combinations of platinum and other metals, recording their electrical and heat characteristics.

HIGH RESISTANCE

Then, in the first quarter of 1879, Edison made a major breakthrough. He realized that the kind of metal he needed in his lamp could not be platinum or anything similar. Platinum offered a very low resistance to electricity. To get it to burn brightly, he would have to pump a lot of electricity through it to create the "electrical friction" inside the burner that created heat. If hundreds—even thousands—of light bulbs were to be made out of platinum and hooked up to a power station, the station's electricity would quickly be drained by these low-resistance light bulbs.

Instead, what was needed was a light bulb whose burner was more resistant to the flow of electricity. Then not very much electricity—current—had to flow through it to create the heat. Instead, the pressure (voltage) of "pushing" the electricity through the resistant wire would create the electrical friction for heat and light. Rather than a lot of electricity, a lot of voltage would be needed, on the order of about one hundred volts per lamp.*

Apparently, this thought had not occurred to most of Edison's rivals. While they each had been working to perfect the light bulb,

* Think of a lot of people trying to get out of a crowded building. If the doorway is wide, offering little resistance, lots of them can flow through easily. It would take a stampede of people to get them "heated up." But if our doorway offered lots of resistance, so that only one or two people could get through at a time, it would take lots of pushing and shoving (voltage) to get the same number of people out the door as before. And in the process, lots of people would get "overheated."

they did not stop to look at the entire lighting system that would have to supply the power. It would be the needs of the entire system that dictated the burner, not the whim of technicians sweating in a laboratory. Adding up the number of homes that needed to be serviced, Edison calculated the specifications for the light bulb. By lengthening and thinning the platinum wires, Edison was able to increase their resistance. By sucking all the air out of the bulb (or as much as his pumps allowed him to), he could extend the time elements glowed before they burned out.

DESIGNING THE WHOLE SYSTEM

While waiting for the right burner to be invented, Edison designed and built the rest of the system: dynamos to produce high-voltage electricity; meters, switches, and fuses to control it. He even hired a glass blower to design special pumps to evacuate the bulbs.

But all to no avail. Despite round-the-clock work by the legions of Menlo Park, by the fall of 1879 it appeared that even the Wizard had met his match. The platinum bulb stubbornly refused to keep burning. High-resistance ones or low-resistance ones, they lasted but a few hours.

CARBON TO THE RESCUE

It was at this point that Edison's research staff turned to carbon. Edison was no stranger to carbon. Menlo Park produced batteries, wires, and other electrical devices out of carbon. Edison's telephone transmitter (what we call the mouthpiece today) was made from carbon.

Carbonizing paper and cardboard had become routine at Menlo Park.* Edison's researchers had experimented with carbon on and

* Carbonizing involves baking a substance in an airless container until only a black, carbon skeleton remains.

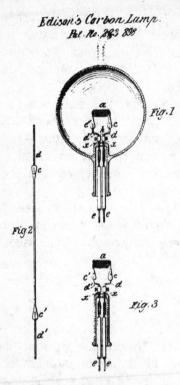

Edison's Carbon Lamp.
Pat. No. 223 898

Fig. 1

Fig. 2

Fig. 3

Fig. 1 = The Lamp in section.
  a ; the carbon spiral or thread.
  cc'; the thickened ends of the spiral.
  dd'; the platina wires.
Fig. 2 = The plastic material before being wound
    into a spiral.
Fig. 3 = The spiral after carbonization.

Edison's Patented Bulb—It Never Worked. Edison's patented light bulb, shown here, never worked well enough to be marketed. Its carbon spiral was too brittle.

off in 1877 and 1879, only to have given up on it after the burners had failed. In addition, many of Edison's competitors had used carbon in their experiments years before he did. A *Scientific American* article of July 1879 mentioned how English inventor Joseph Swan had designed an incandescent lamp using carbon shaped into the form of a cylinder. Edison admitted to having read this article. (Swan defenders claim Edison stole this idea from Swan. Edison backers claim Edison read the article after he designed his own carbon filament.)

In any event, carbon seemed to be the perfect material. Carbon offered the desired electrical resistance. That October Batchelor started the tedious process of finding the right form and shape of the stuff. He tried making carbon spirals, but they crumbled. He tried molding sticks, but they cracked.

Batchelor then turned to carbonized cotton thread and made an important discovery: On October 22, he wrote in his notebook that the carbon-thread "filament"—a thin long piece of carbon—had burned for thirteen and a half hours. A great triumph. In addition, it had a relatively high resistance to electricity. Batchelor was on the right path.

What about other plant materials? Would they work even better? The search was on. Batchelor tried almost anything he could get his hands on: paper, thread, fishing line, cardboard, cotton soaked in tar.

A SUPER VACUUM

Edison sketched out a patent application. His drawing depicted the carbon filament in the shape of a spiral. Filed with the patent office in November 1879, this light bulb design was never successfully put into production. The filament was too brittle. But the patent described a key difference between Edison's light bulb and those of his competitors. Edison's filament would be heated in a bulb containing a super-high vacuum, produced by vacuum pumps Edison had helped to modify to his own standards. By pumping out extra air, Edison was able to slow down the rate at which his carbon filament oxidized (burned up). This achieved longer bulb life.

By New Year's Day 1880 Edison was proudly showing his imperfect lamp to the public. People flocked to Menlo Park to witness carbonized cardboard that burned but was not consumed. Company insiders knew that Edison had not found the best filament material. And it wasn't for lack of trying. They literally scoured the world. Finally, following the Fourth of July holiday, the successful material was unceremoniously announced: bamboo. Bamboo plucked from a common fan was tested and found to burn for an hour and a quarter. It was strong and sturdy enough to stand up to the shaking the filament would take in use.

By August a better quality bamboo, called Japanese bamboo, proved to burn longest. Called the best lamp ever made from a vegetable product, this carbonized bamboo filament burned for almost three and a half hours. It filled the bill for a commercially viable product. And the commotion its discovery caused at Menlo Park? Just another day at the office.

By the end of the year bamboo lamps were burning for 240 or more hours.

THE MAKING OF A MYTH: HYPE AND HOLLYWOOD

According to light bulb mythology—aided years later by Hollywood—there occurred one magical day, October 21, 1879, on which the world stopped turning: the day Edison discovered the secret of the incandescent lamp—carbon. For many years, that day has been celebrated as Electric Light Day by power companies, and its fiftieth anniversary, October 21, 1929, was chosen to honor Edison at Light's Golden Jubilee.

How this almost magical story of Edison hitting upon carbon as the answer to his problems got started is not difficult to piece together. It's not hard to imagine how an eager public—aided by erroneous press reports, romanticized notions of the inventor, and sketchy evidence—could create the aura of revelation akin to Moses on Mount Sinai.

When Edison declared in the latter part of 1878 that he had "struck a big bonanza," he made sure everyone heard him—especially his competitors and potential backers. A story in the *New York Sun* on September 16 carried a headline proclaiming "Edison's Newest Marvel. Sending Cheap Light, Heat, and Power by Electricity." The paper quoted Edison as saying "I have it now," despite Edison having said nothing of the sort.

That self-serving quote was picked up by other newspapers

around the country and had a great impact on the commercial centers of Chicago and Philadelphia. Businesses were lining up to invest. The public was psyched; Edison Company stock sky-rocketed on Wall Street. Edison was on center stage once again, and the whole world was primed for a breakthrough just waiting to happen. As in all good stories, all that was needed was that legendary "magical moment of discovery." That moment was long in coming but arrived neatly packaged just before Christmas as a full-page story in the December 21, 1879, edition of the *New York Herald*.

With headlines screaming "Edison's Light," reporter Marshall Fox shouted to the world his scoop: the inside story of how Edison had succeeded in producing "the perfected lamp."

The Secrets of Success. A full-page spread of Edison's triumph, as depicted by the *New York Daily*.

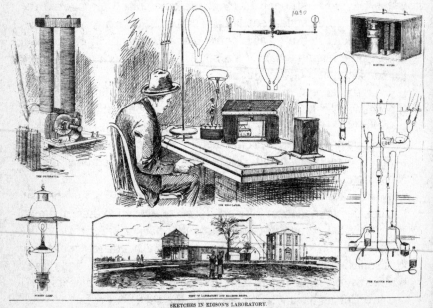

SKETCHES IN EDISON'S LABORATORY.
ILLUSTRATING THE MECHANISM EMPLOYED IN PRODUCING HIS ELECTRIC LIGHT.

Fox went on to explain how in October the platinum filament lamp was ready to be unveiled to the public except for a "few minor details" (one of them being that the filament was not satisfactory). Then, according to Fox, there occurred a discovery that changed the whole course of light bulb history:

> Sitting one night in his laboratory reflecting on some of the unfinished details, Edison began abstractedly rolling between his fingers a piece of compressed lampblack until it had become a slender filament. Happening to glance at it the idea occurred to him that it might give good results as a burner if made incandescent. A few minutes later the experiment was tried, and to the inventor's gratification, satisfactory, although not surprising, results were obtained.

This light bulb supposedly burned for forty hours. What a great story! Except for one ugly detail: There is no record of this legendary light bulb. Despite Edison being quoted as saying that this account of the story is essentially true, records of the laboratory show nothing of great interest happening on October 21. The closest one comes to a great "breakthrough" occurred the next day when Batchelor's carbonized cotton glowed for a few hours. But what about the popular accounts of Edison thoughtfully contemplating a piece of lampblack and absentmindedly rolling it into a "slender filament"? The stuff of Hollywood (remember Mickey Rooney?).

People tend to romanticize what they can't quite remember. And Edison's "breakthrough" is probably the result of the interested parties reminiscing years later over the tedious process of trial and error so common in invention.

SWAN: A CLOSE SECOND

Edison's patent would be followed by years of court battles over who deserved to be credited with the invention of the light bulb. In the strict sense of the meaning of the word invention, Edison can hardly claim to be the bulb's sole inventor. His invention contained no unique materials. He merely improved upon the work of others.

On New Year's Day 1880, two months after Edison's famous "event," a letter appeared in the British journal *Nature*, submitted by Joseph Swan and alluding to the hullabaloo about Edison. What's the big deal, he wrote:

> Fifteen years ago, I used charred paper and card in the construction of an electric lamp on the incandescent principle. I used it, too, in the shape of a horse-shoe, precisely, as you say, Mr. Edison is now using it.

Swan was England's Edison. Or to put it more correctly, Edison was America's Swan. A tinkerer and born inventor, Swan not only showed great interest in the light bulb but made key improvements in the science of photography. Twenty years Edison's senior, Swan was experimenting with light bulbs while Edison was still a toddler. By 1848, at the age of twenty, Swan had already begun experimenting with a carbon/vacuum light bulb. And by 1877 he turned his full attention to mastering the light bulb. His publication in *Scientific American* of experiments with a carbon cylinder was certainly read by Edison. But there is no reason to believe that Edison's key light bulb breakthrough—making the carbon into a thin filament, not a rod—was influenced by Swan's work. If anything, Swan swiped the filament idea from Edison. Swan never claimed to have invented the carbon filament.

Having installed hundreds of light bulbs in some of England's most prestigious homes (Sir William Thompson's home glowed with 150 bulbs) and landmarks (the Royal Institution), Swan started his own company. This caught Edison's attention—more correctly, his lawyers' attention. In 1881 they went into English court to sue Swan for patent infringement. Fighting on his own turf, Swan proved to be a formidable enemy. Bowing to good business sense, Edison couldn't lick 'em, so he joined 'em. A new company, Edison-Swan United, was born and sold light bulbs under the trade name

Ediswan. Together, the transcontinental giants put all light bulb competitors out of business.

THE BIRTH OF THE GREAT WHITE WAY

Even after the spectacular triumph of the electric light bulb, the world did not rush to embrace this new technology. Remember, no power lines, no house wiring, no power plants existed. All this took time to come on line. In 1885 the British *Electrician* considered the electric light "at this moment a luxury." Gas was in no danger of being replaced. Only the super-rich could afford the shiny glass bulb. In Europe the incandescent lamp numbered among its elite environs the chateau of the Baron Alphonse de Rothschild and the palace of Prince Manko-Negoro. The castle of the Marquise of Bute was illuminated by the glow of 400 electric lamps. In England, 150 Swan lamps glowed in Didlington Hall, Brandon; C. P. Huntington's mansion at Astely Bank glowed in the light of the new lamps.

In the United States, Macy's in New York City became the first store to use incandescent lamp lighting in 1883, after having installed arc lights in 1879. In 1888 the Hotel Everett on Park Row in New York City became the first American hotel to be lighted by incandescent lamps. Its main dining room, reading rooms, parlors, and lobby shone brightly under the glow of 101 electric lights.

New York theaters were the first to take advantage of the lights' ability to spell out "boldly in letters of fire" the names of the featured program. The Great White Way began to go incandescent in 1895.

By 1882 Edison Electric Light Company's Pearl Street Station was up and running in New York City. This early electrical system, running on direct current (see Chapter 3, "The War of the Currents"), would serve as a model for the rest of the country.*

---

* Ironically, Edison's power grid, which relied on direct current, would not last. His battle with Westinghouse over New York's electrical system would spell the end of DC and lead to the modern system of AC house current.

THE LIGHT BULB HALL OF FAME

Date	Inventor	Nationality	Filament	Atmosphere
1802	Davy	English	platinum	air
1840	Grove	English	platinum	air
1841	De Moleyns	English	platinum	vacuum
1845	Starr	American	platinum	air
			carbon	vacuum
1846	Greener	English	carbon	air
1848	Staite	English	platinum	air
1850	Shepard	American	carbon	vacuum
1852	Roberts	English	carbon	vacuum
1856	de Changy	French	platinum	air
1859	Farmer	American	platinum	air
1860	Swan	English	carbon	vacuum
1872	Lodyguine	Russian	carbon	nitrogen
1875	Woodward	Canadian	carbon	N/A
	Kosloff	Russian	carbon	nitrogen
	Konn	Russian	carbon	vacuum
1876	Fontaine	French	carbon	vacuum
1877	Maxim	American	platinum	air
1878	Sawyer	American	carbon	nitrogen
	Maxim	American	carbon	hydrocarbon
	Lane-Fox	English	platinum-iridium	air-nitrogen
	Farmer	American	carbon	nitrogen
1879	Jenkins	American	platinum	air
	Hall	American	platinum	air
	Edison	American	carbon	vacuum

If you think Edison was the only—or the first—experimenter working on perfecting the incandescent light bulb, check out the *partial* list above. Look how many inventors had already tried using carbon prior to Edison. (Adapted from *The Chronological History of Incandescent Lamps*, General Electric Incandescent Lamp Historical Committee, February 1922.)

The War of the Currents, or Let's "Westinghouse" Him

An awful spectacle, far worse than hanging.

>—*New York Times,*
>August 7, 1890

It has been a brutal affair. They could have done it better with an axe.

>—GEORGE WESTINGHOUSE,
>August 7, 1890

The grandest success of the age.

>—ALFRED SOUTHWICK, New York
>State Commission on Humane
>Executions, August 16, 1890

ON AUGUST 6, 1890, William Kemmler became the first man to die in the electric chair. A twenty-eight-year-old fruit vendor and resident of Buffalo, Kemmler murdered his girlfriend Tillie Ziegler with the blunt end of an axe. He was sentenced to die at Auburn State Prison by a new and supposedly painless method: "the application of electricity." A drunk who spent most of his time dazed in an alcoholic fog, Kemmler met a death that was anything but painless. When the executioner pulled the switch, Kemmler refused to die.

27

The newfangled electric chair did not send enough electricity to his body. Only after repeated shocks did the victim succumb to a gruesome death that received anything but praise from its horrified observers. "The guillotine is better than the gallows, the gallows is better than electrical execution," is how Dr. E. C. Spitzka, attending physician and witness, phrased it.

It was a nightmare not only for Kemmler but for the state of New York. And a black eye for the state of electrical innovation. Because as it turned out, Kemmler was more than a victim of his crime; he was a pawn. The electric chair—and Kemmler's death—were merely the byproducts of a vicious battle between two giants of the industry, Thomas Edison and George Westinghouse, to control the future of electric generation.

THE WAR OF THE CURRENTS

The struggle for the future of electricity became known as the War of the Currents. Staged over a hundred years ago in New York, it pitted Edison against fellow tycoon Westinghouse in a fight involving which type of electricity would become standard: direct or alternating current (DC or AC).

Edison had built his reputation and business on DC. His famous light bulb and his entire electrical system ran on electricity that traveled in only one direction, that is, direct current. Edison had just installed his Edison Electric Light Company in downtown New York in 1882 to sell DC electricity to power his bulbs, electric cable to carry the power, and electrical generating devices to make it. Edison had established a powerful grip on the largest city in the country and was not ready to surrender his hard-won turf (he had just shown how electricity was superior to gas lighting) to any competitor.

But George Westinghouse had other ideas. His Westinghouse Electric's alternating current (AC) system was far superior to Edison's. Being just a year older than Edison, this young tycoon hoped to do to Edison what Edison had done to the gaslight industry: try to put it out of business. But Westinghouse faced a seemingly impossible challenge in trying to shove aside the ego—and power—of Thomas Edison.

Edison controlled a virtual monopoly in the direct current electrical business. When he went after his famous light bulb Edison didn't merely invent a better bulb. He designed the entire system to go with it: power generating stations, power lines, and motors. Edison's notoriety, his invention of the "speaking phonograph" just twelve years before, made him an icon of American folklore at the grand old age of forty-two. Moving out of his cramped quarters at Menlo Park, Edison had set up a new kingdom worthy of his reign in West Orange, N.J., a new laboratory ten times Menlo Park's size.

By the end of 1887 a total of 121 Edison central power stations were operating or under construction. Taking on Edison at his own game was like Toyota taking on GM in Detroit.

EDISON VS. WESTINGHOUSE

But Westinghouse was just the man to do it. A self-made multimillionaire who whirled around the country in his private Pullman train, the inventor of the railroad airbrake was Edison's equal. While the Wizard of Menlo Park was a disheveled loner, Westinghouse was fun-loving and stylish. But both possessed an iron will that could not stand second place.

Westinghouse had already investigated and invested in alternating current. AC was being used as an acceptable means of producing powerful lighting; its bright electric arc lights lit up streets and movie projectors. By 1885 he had bought the patent rights to a new electrical invention called the transformer. With it, he held the means of delivering electricity to distant homes and factories in a way that Edison's DC could not.

Edison's DC generators produced the same 120-volt "house current" that was fed to a consumer's electrical socket (later his three-wire system would produce 240 volts DC). But because DC loses strength in the wires, DC electrical generating plants had to be built close to customers. That meant lots of them spread around the city, perhaps even in the backyard of each homeowner or apartment building.

Westinghouse's newly acquired AC transformers changed all that. AC possessed a property that DC did not have: AC could be—via the transformer—"stepped up" to thousands of volts at the power plant. The power plant could then be positioned many miles outside town and the electricity shipped by wire to homes and factories. Who cared if it lost a few volts along the way? At a power pole outside a home or building another transformer waited to "step it down" back to 110-volt house current.

This meant that a central power plant could be located anywhere in or out of town near a coal supply or the water power needed to run it. Small towns, too poor to pay for costly electric plants scattered around their cities could build them out of town and share the costs with other small cities. The device would transform more than just voltage: It would make electricity a cheap and plentiful utility.

In 1886 Westinghouse opened the first successful AC system in America in Buffalo, N.Y. A year later, after buying up small companies and electrical patents, he was operating more than thirty such plants, mostly in the South and West. By the end of 1887 he began to penetrate into other DC markets supplying electricity to almost 135,000 light bulbs.

But the plum—New York City—eluded him. Edison had wrapped up New York in his DC package and Westinghouse lacked the one ingredient needed to challenge the Wizard on his own turf—an AC electric motor. Who would buy electric power limited to lighting up a bulb while Edison's electric was making motors hum?

Edison watched. As far back as 1880 he had decided that DC was the only way to go. European engineers had offered in 1886 to sell him an AC transformer system but he turned them down. He believed that alternating current was costly and dangerous to work with. Edison's colleagues argued that while 1,000 volts of AC would "kill a horse," the safety of 240 volts DC was "undoubtedly absolute." His aging peers counseled him against AC: great electrical scientists of his time, Europeans like Lord Kelvin and Werner von Siemens and Americans like Elihu Thomas (who later changed his mind) and Franklin Pope.[1]

Did Edison truly believe AC current to be too dangerous to tame—even by the talents of an Edison—or was he becoming too careful in his aging years? Perhaps the man who once relished the idea of single-mindedly experimenting the problems out of a system had become old and conservative. Could this be the same man who had spent so many fruitless hours locked up with a light bulb?

Edison realized that to stay with DC meant a power station literally in everyone's backyard. DC power could only travel a couple of miles before being turned to heat in the wires. Turn-of-the-century technology did not allow him to step up or step down DC voltage. But Edison was, if nothing else, bullheaded. He would defend DC power with all his might.

WARNING!

When it appeared that AC was making significant inroads into Edison Electric's subsidiary's territories, Edison struck back with a scathing propaganda campaign. In February 1888 Edison showered West Orange with red leaflets screaming A WARNING! FROM THE EDISON ELECTRIC LIGHT CO. Each eighty-three-page leaflet accused Edison Electric's competitors of being liars and thieves. The leaflet was intended to scare people. It warned of the dangers of high voltage to those who would come near high-tension AC power lines. It listed unfortunate souls who had fried to death in theater and factory accidents. Is this the kind of electricity you want to be using in your home? Any responsible parent would have to say no.

Edison Electric also appealed to the fears of corporate America, assuring the "baby Edison" electric companies they had little to fear from the "inevitable deficiencies of the alternating system."

The scientific community stepped in as a referee. The Chicago Electric Club conducted highly visible and well-publicized debates comparing the two.

But then later in that year, 1888, things turned ugly. No one can say for sure why, in Edison's mind, Westinghouse's AC system suddenly switched from having "inevitable deficiencies" to becoming a real threat. How could a system barely a year old grow so quickly to rival Edison's entrenched power base? Perhaps it was all a matter of good timing.

In 1887 the price of copper began to rise sharply. Shortly after a French syndicate cornered the copper market, it drove the price of copper up from nine cents per pound to twenty cents per pound. DC systems relied heavily on fat copper wires to carry the current. High copper prices meant the cost of installing DC systems would skyrocket. On the other hand, high voltage (two thousand volts) alternating current could still be carried on skinny copper wires. And perhaps in comparing the operating costs of the two systems, Edison now for the first time perceived AC to be a real threat.

Perhaps.

But the crucial development that changed the complexion of this conflict was the trump card George Westinghouse was about to play in the summer of '88 with the help of a brilliant Croatian immigrant, Nikola Tesla.

TESLA TO THE RESCUE

Entire books have been written about the quirky genius who had once worked for Edison. Tesla was a man who held hundreds of patents and would become a legend in his own time. Nikola Tesla's brilliance came to light under the watchful eye of George Westinghouse. Tesla held the key to Westinghouse's success in battling Edison; in April 1888 Tesla announced he had invented the AC motor.

Seeing the elegance of the idea, Westinghouse immediately bought the rights to Tesla's motor (and to one made by Galileo Ferraris in May of that year). Thinking on a grand scale, Westinghouse hired Tesla to design an entire AC power grid for a large city. The railroad magnate had decided it was time to take on the Big Apple.

Nikola Tesla. The unconventional genius seen sitting among "lighting bolts" of electricity in his lab. Tesla was invaluable in helping Westinghouse defeat Edison's bid to control the electrification of New York City.

Now Edison Electric faced a double-edged sword: the high cost of copper wiring and the superiority of the AC system. And when it appeared that Tesla's AC motor would pose a real threat to Edison, he embarked on a campaign to get AC outlawed. Failing that, he'd make sure no one in their right mind would use it. Edison engaged in a long and torturous mudslinging battle against Westinghouse. If economics and scientific reasoning could not stop alternating current, perhaps nature's most basic emotion—fear—could.

And Edison had no trouble finding allies eager to do the dirty work.

His first opportunity showed itself in the person of a "Professor" H. P. Brown, a self-employed, self-promoting electrical consultant. In the early days of the War of the Currents, Brown had written an open letter in the *New York Evening Post* that appeared on June 5, 1888. In it, Brown asked that alternating current be outlawed because it was "damnable" and "dangerous." On the strength of that letter, Edison put Brown to work in his laboratory.

Brown grew quickly to become a hard, fast friend of Edison and his point man in the battle. An inventor of DC power arc lamps, Brown claimed that AC was intrinsically more dangerous than DC even at low voltages. Since high-voltage wires, with telephone and telegraph cables, literally blackened the skies of New York with their numbers, Brown had an attentive audience. New Yorkers were no strangers to the dangers of high-tension wires. Falling cables killed scores of horses and pedestrians each year.

Professor Brown claimed that it wasn't the strength of the electricity that did them in, but the nature of it—namely AC. Direct current, he argued, was safer even at high voltages. He called for the outlawing of all high-voltage (above three hundred volts) lines from the city. Brown knew that AC lost its advantage over DC at such a low voltage. Such a law was bound to favor DC.

To Westinghouse this was preposterous. Who was this guy? How could he make such outrageous claims? Didn't people realize that Brown held patents for DC equipment? He was just a tool of the DC industry.

To answer these charges Edison set up Brown as a lab assistant, giving him free rein to prove AC's lethality and the services of his top engineer, Arthur Kennelly. Brown was to fire the first deadly salvo in the War of the Currents. Its first casualties were to become the dogs and cats of West Orange, New Jersey.

SHOCKING THE NEIGHBORS' PETS

In the summer of 1888 Brown and Kennelly paid the neighborhood children of West Orange twenty-five cents for each pet they could bring in. Then they conducted midnight experiments—to avoid detection by the ASPCA—in which they shocked the animals to death with alternating current.

On July 12 Kennelly wrapped wet bandages, affixed with copper wires, around the paws of a thirteen-pound fox terrier. Starting with four hundred volts DC, a series of shocks were applied to the animal until it finally died. Kennelly and his colleagues kept records of the amount of voltage, time applied, and apparent state of the dog. They claimed that edging a little dog onto a sheet of tin and shocking it to death with a jolt of one thousand volts AC was actual "scientific" proof of the dangers of alternating current.

Two days later, electrodes were fixed to the skull of a German shepherd and direct current passed between its ears. Kennelly noted, "While the current was passing there was no struggling or yelping, and beyond the half closing eyes and a trembling in the head, it would be difficult to see the effect of the current."

All told, Kennelly and Brown decimated the pet population of West Orange, in the process crudely killing fifty dogs and cats in these "experiments."

They staged special public animal electrocutions for the press at Columbia College's School of Mines in New York. Edison supplied the equipment. Reporters were invited to one memorable execution, July 30, 1888, to witness a "practical demonstration." Hardly ready for the ghastly exhibition they were about to see, the journalists gasped as two of Brown's assistants hauled in a snarling, uncooperative seventy-six-pound Newfoundland dog, set it in a wire cage, and turned on the juice. SPCA delegates arrived too late to save the dog, who died after repeated jolts of three hundred, four hundred, five hundred, seven hundred, and one thousand volts DC and a final zap of three hundred volts AC.

In the greatest tradition of "a little knowledge is a dangerous thing," Brown challenged Westinghouse to an electrical duel. Throwing down the gauntlet in the *New York Times* of December 18, 1888, Brown dared Westinghouse to a dangerous contest:

> I . . . challenge Mr. Westinghouse to . . . take through his body the alternating current while I take through mine a continuous current. . . . We will begin with 100 volts and will gradually increase the pressure 50 volts at a time . . . until either one or the other has cried enough, and publicly admits his error.

Westinghouse ignored the stunt.

Finally, Brown went for the jugular. He would make AC so

unpalatable as to be fit not only for executing dogs and cats but people. Could someone on death row be in need of a public execution?

THE WORLD'S FIRST EXECUTION BY ELECTROCUTION

Enter the state of New York. The Empire State had been looking for a more humane way of execution to replace hanging. The state had seriously mishandled several hangings in the early 1880s. Sometimes the hangman's noose had been too loose, resulting in prisoners slowly strangling to death. In other cases the rope had been too tight, decapitating the prisoners in a gory display of capital punishment. Having considered and rejected three dozen other methods, Governor David B. Hill appointed a committee to study the use of electricity as an instrument of electrocution.

After initially turning down the committee's request for an endorsement,* Edison put the considerable weight of his prestige behind the idea. He felt electricity would "perform its work in the shortest space of time, and inflict the least amount of suffering upon its victim." And he especially singled out the type of electric generator to be used as "alternating machines, manufactured principally in this country by Geo. Westinghouse . . . The passage of the current from these machines through the human body, even by the slightest contacts, produces instantaneous death."[2]

Edison's opinion greatly influenced the committee's decision to recommend electrocution. Its chairman, Elbridge Gerry, testified later that he considered Edison to be "somewhat of an oracle," leaving "no doubt after hearing his statement of it."[3] On June 4, 1888, Governor Hill signed a bill that made the electric chair replace the noose. Electrocution would be the execution of choice in New York State as of New Year's Day, 1889. But how the electrocution should be conducted—using what kind of electricity, how it should be applied, and for how long—was not specified in the law. Finding

---

* Edison was opposed to capital punishment.

this out would be the task of New York's Medico-Legal Society.

Brown, who by now was the world's leading authority on electrical execution of four-legged prisoners, made it his job to see that the first human executed in New York would be fried by AC current. If Brown couldn't get AC outlawed, he'd get it equated with murder.

Hired as a consultant to the Medico-Legal Society by a colleague on the committee, Brown decided to prove the ability of AC to electrocute large human-sized animals as well as small ones. Skeptics had voiced opposition to electrocution not out of humanitarian considerations but practical ones: After all, people were a lot bigger than dogs.

ELECTROCUTING FARM ANIMALS

On December 5, 1888, Brown hauled in two healthy calves and a 1,230-pound horse to the Edison dynamo room and zapped them right there with a lethal thirty-second jolt of AC. "In fifteen seconds," reported the *New York Herald* about one of the calves, "the victim was veal." See? No problem with human-sized mammals. Humans should succumb as quickly as these animals. Brown and Kennelly told the newspapers that the results could not have been more satisfying.

That gory display, coupled with Edison's testimony, was enough to convince the state. Less than a week later, on December 10, Brown's unending lobbying paid off. The committee voted unanimously to adopt alternating current as its means of electrocution. New York became the first state to sanction electrocutions carried out, of course, using alternating current.

But first Harold Brown had to find an AC generator. The only one who made them was Westinghouse, and he wasn't about to sell one to Brown. By now Westinghouse was past the point at being shocked at Brown and Edison's tactics and distortions of the truth. Westinghouse strongly protested the use of any of his generators

Experiments in killing animals by alternating current in the Edison Laboratory at Orange, N.J., December 5, 1888.

to electrocute a prisoner. But Brown was able to buy three of them, deviously, by means of a third party, and had one shipped back to the prison in Auburn.

STAY OF EXECUTION

Lawyer Bourke Cockran, defense attorney for the condemned William Kemmler, won a stay of execution on the grounds that electrocution was cruel and unusual punishment and thus was prohibited by the New York State Constitution. Called as a witness to provide expert testimony about the correct way to electrocute a person, Edison, as usual, was not at a loss for words. Under heavy cross-examination, he admitted knowing hardly anything about the human anatomy or the effects of electricity on it.[4]

COCKRAN: You do not claim to understand anything about the
          structure of the human body.
EDISON: No, sir; only generally.

Was blood or muscle a better conductor of electricity? Edison
said he thought blood was the better conductor but he'd know
better after conducting experiments.

COCKRAN: Do you know anything about the conductivity of
          the brain?
EDISON: No, sir.

Then why was Edison so sure that electrocution was such a
painless way to die? Edison replied that he based that belief on what
he read, on what other people had observed, and from seeing "one
or two" animal electrocution experiments conducted by his employees.

Yet despite Edison's admitted ignorance of the effects of electricity on the body, his lack of scientific proof to back his claims, it
was the opinion of the newspapers covering the case that the mere
presence of Edison at the dock, giving expert testimony even of
questionable accuracy, was enough to sway the judge. As the
Albany Journal put it:

> The Kemmler case at last has an expert that knows something
> concerning electricity. Mr. Edison is probably the best informed man
> in America, if not in the world, regarding electrical currents and their
> destructive powers.[5]

Pointing out that there was no better authority than Edison on
the subject of electricity, the *Wilkes-Barre News Dealer* told its
readers on July 25, 1889, that Edison's testimony would show
there was "no doubt as to the efficacy of electricity as a death-
dealing agency."

No wonder, then, that the judge denied Kemmler's appeal and
ordered him electrocuted. As the *New York Times* would point out

a year later, "It was largely due probably to such testimony . . . that the law was upheld finally in the courts."

THE FIRST USE OF "THE CHAIR"

Finally, at an ostensibly secret ceremony on August 6, 1890, William Kemmler was seated in the electric chair and the switch was thrown. But the victim refused to die. For seventeen seconds electricity passed through Kemmler's body, but when the juice was turned off the prisoner was still alive. "The body remained limp and motionless," noted physician Dr. Carlos F. MacDonald, "when there occurred a series of slight spasmodic movements of the chest" as if the prisoner had tried to breathe.

Let's "Westinghouse" Him

Trying to add insult to injury, Edison and his colleagues began to search for a word that would exactly describe the fate awaiting William Kemmler on death row. In 1889 the word electrocution had not become an accepted term. Edison suggested "ampermort," "dynamort," and "electromort." At the request of one of Edison's colleagues, Eugene Lewis, an attorney suggested "electricide" and went on to make an even more macabre suggestion:

"There is one other word which I think . . . might be used with some propriety. It can be used as a verb and as a noun to express kindred ideas. The word is 'westinghouse.' As Westinghouse's dynamo is going to be used for the purpose of executing criminals, why not give him the benefit of this fact in the minds of the public and speak hereafter of a criminal as being 'westinghoused'; or to use the noun, we could say that . . . a man was condemned to the westinghouse. It will be a subtle compliment to the public services of this distinguished man. There is precedent for it too, one that could not be more apt or appropriate. We speak of a criminal in France as being guillotined. . . . Each time the word is used it tends to perpetuate the memory and services of Dr. Guillotine, who afterwards died by the same machine that he invented."

The Edison Electric Chair. Used in the 1892 execution of Charles MacElvaine. After the prisoner's hands were strapped into buckets of salt water, a charge of sixteen hundred volts applied through the hands for fifty seconds failed to kill the victim, who made gasping and coughing sounds. A second jolt applied to the backup head and calf electrodes resulted in death. Edison, who had no expertise in human anatomy, refused to admit his design was faulty. His chair was never used again. All future executions relied on the head-to-calf electrode.

A second shock was applied, and this time the AC current was kept on for seventy-two seconds. Only after a "small volume of vapor, and . . . smoke was seen" was the electricity shut off. The event would be described by the *New York Times* as "an awful spectacle, far worse than hanging."

THE WAR IS OVER

Despite the awful spectacle, the long and malicious effort to denigrate alternating current came to naught. Even the electric chair couldn't kill it. In the War of the Currents, Westinghouse won big. His AC system became widely accepted. The merits of AC over DC were obvious to most electrical engineers. The debate came to a close when Westinghouse created a great sensation at the Chicago World's Fair of 1893, where a contract to light the entire fair using AC was awarded to him. As icing on the cake, he even showed off an AC motor that could convert DC to AC. Now he could swallow up any remaining Edison DC systems into his.

Westinghouse's ultimate victory came in the fall of 1893. A contract to build the world's first hydroelectric power project in Niagara Falls, N.Y., was awarded to him over bids by Edison. There, in 1895, three five-hundred-horsepower Westinghouse AC generators drew power from the thundering river and sent it twenty-five miles over power lines to Buffalo.

It is more than slightly ironic that the electrical cables carrying that alternating current were built by General Electric, a company formed out of the merger of Edison's electric company with Thomas-Houston, the second biggest company (next to Westinghouse) in the AC business. Edison had seen the light.

Edison gave up the chase and finally retreated to his laboratory in West Orange. Putting himself back into productive work, he went on to develop motion pictures, improve the phonograph, and create the kingdom called Edison General Electric.

As for Brown, his foray into electrocution lasted a bit longer. Trying to fine-tune the electric chair, he shocked a few horses to death in Sing Sing. In an example of poetic justice, it was Brown whose name became synonymous with the "death current" at the time.

EPILOGUE

Did Edison really believe AC to be as evil as he and his hired guns fashioned it? In his letters, Edison warned colleagues of its lethality. Edison did believe AC to be highly dangerous. But altruism aside, he worried about business. He wrote that if AC shocked to death too many people, the fledgling electric generating industry would suffer.

What about his claims that AC was more dangerous than DC? Tests today bear out this conclusion. Comparing DC to AC house current, it takes three times more DC than AC (sixty) to produce cardiac arrest.

POSTSCRIPT: BLINDED BY THE LIGHT

If Edison had not been so stubborn, he might have paid closer attention to the mysterious effects alternating current produced on his light bulb. This mystery eventually led to the invention of the vacuum tube. But Edison's almost religious devotion to direct current as a means of carrying electric power precluded him from becoming father of the radio revolution.

Edison noticed that after his new carbon light bulbs glowed for a while, the glass on the inside of the bulb would turn black (you see it to a much lesser degree on modern bulbs). Thinking this phenomenon worthy of investigation, he took a closer look. This black stuff was blocking the light and must be done away with. But how?

Edison concluded that the carbon must be boiling off the negative side of the filament since it was that side that tended to burn out. No problem here. But something else bothered him: While the carbon evenly coated the glass envelope, the carbon left a clear line—a "shadow"—on the glass near the positive side of the filament. Why?

The positive side must be attracting the carbon, he reasoned, which meant that the carbon particles must carry a negative charge. Testing his hunch with an experiment, Edison found that if he placed a wire in the bulb (without touching the filament) and attached the other end to the positive electrode of the battery, he could measure a flow of electricity. But if he switched electrodes and attached the wire to the negative battery terminal, no electricity would flow. The results backed up his hypothesis that the current carriers were negative charges. Very intriguing.

Next he substituted a flat metal plate for the wire. The one-way current flow was even more intense. As he varied the brightness of the filament, Edison noted that the current flow varied in step with the change in applied DC voltage. Turning this discovery into a marketable item, Edison patented a modified version of this device in 1883 as a means of measuring voltage. He then lost interest in the phenomenon and turned to other things.

Had he been a bit more open-minded about alternating current, Edison might have made a dramatic discovery. Had he hooked up an AC voltage to his wire or flat-plate electrode, he would have noticed that current flowed at only half the time, that is, in one direction. He would have discovered a way of changing bi-directional AC into one-way DC. But his abhorrence of AC clouded his thinking. Years later, in 1884, William Preece, Chief Engineer for the British Post Office, would come to Philadelphia to attend the International Electronic Exhibition. He would see Edison's light bulbs on display, meet the inventor, and go home with one of these newly modified tubes with the unusual metal-plate electrode.

Preece would experiment with the bulbs and coin the term "Edison effect" to describe boiling off of negatively charged parti-cles from a hot filament. Ironically, a consultant to the Edison Electric Light Company of London, John Ambrose Fleming, would confirm the results but with a significant additional finding: Electric current also flowed when an AC voltage was applied to the filament, and the current that came out of the bulb was no longer AC but had been converted to DC.

Fleming publicly reported these findings in 1896. This discovery led years later to the invention of the vacuum tube radio and opened the door to modern electronics. And it all happened because some-one besides Edison decided to investigate AC current in the light bulb. Fleming would eventually become "Sir" Ambrose, knighted by King George V almost half a century after his pioneering work.

4

....

You Press the Button, We Do the Rest

*My layout, which included only the essentials, had in it a
camera about the size of a soap box, a tripod which was strong
and heavy enough to support a bungalow, a big plate holder, a
dark tent, a nitrate bath, and a container for water.*

—GEORGE EASTMAN

IF YOU HAD to lug around a darkroom, the number of pictures you'd
take on vacation would fall drastically. There's no fun in setting up
shop in Disney World to shoot, develop, and dry your photos while
waiting in line to ride Dumbo. But that's what photographers had to
do if they wanted to take pictures a hundred years ago. They
literally had to carry their darkrooms on their backs. In addition to
the twenty- to thirty-pound camera, photographers in the early
days of photography (back in the 1870s) carried a heavy tripod,
gallons of noxious photographic liquids, and heavy glass plates.

Simple family portraits, incomparable pictures of the Civil War,
and all the scenes of the mid-nineteenth century were taken the
same way. A professional photographer would set up a darkroom
on the spot—it could be in your home, it could be on the Get-
tysburg battlefield. If you were sitting for the photo the photographer would go into the darkroom, open up the bottles of evil-
smelling chemicals, mix the emulsion, and spread it on a glass plate.

The plate would be slid into a huge camera sitting atop a heavy
wooden tripod (if you were lucky, nothing spilled on the carpet).
The photographer would then drape a dark cloth over his head and
from this blackened tent would ask you to remain motionless for at

45

least half a minute. (Try that sometime!) After the exposure was finished, he would retire back to the darkroom, plate in hand, and if the temperature of the developer and hypo were just right and nothing had contaminated the chemicals, he would emerge triumphant with a usable negative.

The whole process could take hours. And you'd never want to have your picture taken again. But that's the only way photos could be taken in those days; take 'em or leave 'em. You had to have a strong back and incredible patience. But fortunately for all of us shutterbugs, one young and brilliant bank clerk in Rochester, N.Y., decided to do something about it.

THE VACATION THAT CHANGED PHOTOGRAPHY

George Eastman was about to take his first vacation from the bank. At the age of twenty-four, George was excited about traveling to the Caribbean and photographing the crystal-clear waters and beautiful beaches. As a newly interested amateur photographer, George would have loved to have some pictures of his vacation but he shuddered at the idea of lugging along all that equipment. Wouldn't be much of a vacation.

Had George been destined to remain with the bank, our story would end right here. But George was one of those guys who'd rather strike a match than curse the darkness, so he said, "There's got to be a better way; let's find it."

Eastman canceled his vacation and set out to find a better way. He set up shop in his mother's kitchen and worked long nights after a full day at the bank. Mornings found him curled up in a blanket next to the stove. For three years Eastman toiled in his mother's kitchen until he emerged with a new, more sensitive film emulsion.

The next and most obvious target, of course, was the unwieldy camera. But what to do? What made the camera so bulky were the heavy glass photographic plates that were not only a pain in the back but highly fragile: Drop one and you were left with a pile of glass slivers. What if you could do away with the plates? Then the camera could be made smaller, lighter, and easier to take along. But what to take their place? A material that was light and bendable; something that could be rolled up. Paper was the answer.*

Eastman turned to paper and immediately hit a roadblock. Professional photographers did not like his paper-based film. They claimed they could see the grain of the paper on the finished photograph. No thank you; they'd rather stick to the bulky glass plates. Here was a real problem. Without the professionals to buy his new film, Eastman's market would dry up—and along with it his newly formed Eastman Dry Plate Company. Heck with them, thought George. Lots of average folks out there like me would love to have a simple and effective way of taking photographs. Let's shoot for the amateur photographer. The average guy in the street. I'll create my own market. My idea is to "make the camera as convenient as the pencil."

Now all Eastman had to do was figure a way of reducing the size of the camera and making it a cinch to use. His first step was to take the burden of developing the photos out of the hands of the amateurs. He decided to roll the film onto a spindle and stuff the roll into a small light-proof box. Then he placed a small lens on the outside of the box, stuck a "key" on the box to turn the film so that it could be unrolled in the dark, and placed a little red window on the back so the number of exposures could be read.

To take a picture, all one had to do was to wind the film with the key to the point where the number one showed through the red window, indicating when the emulsion-coated celluloid was opposite the lens. You cocked the shutter by pulling a string and pushed a button to take a picture. When you pressed the button, the shutter would expose the film to the light for a split second. After

---

* Not everyone wanted to modernize picture taking. Some romantics, like the one who wrote a letter to the *British Journal of Photography* in 1880, lamented even early attempts at streamlining photography: "This newfangled idea of ready-made plates takes all the fun out of photography. The next stage might be a shop to produce prints and lantern slides to order—but that is too distressing to anticipate."

✳═THE═✳

K O D A K

CAMERA.

Silver Medal at Minneapolis Convention
P. A. of A. for most important invention
of the year.

PHOTOGRAPHY REDUCED TO THREE MOTIONS.

And so on
for 100
Pictures.

1. Pull the Cord.   2. Turn the Key.   3. Press the Button.

✱ ANYBODY CAN USE IT. ✱

Size of Camera, 3¼ x 3¾ x 6½ inches.

Weight, 1 lb. 10 oz.

Size of Picture, 2½ in. diameter.

Uncapping for Time Exposures.

➤➤ PRICE, $25.00 ◄◄

Price includes hand-sewed sole leather Carrying Case,
with shoulder strap and film for 100 exposures.

➤➤ PRICE ◄◄

For Developing, Printing and Mounting 100 Pictures,
including spool 100 films for reloading Camera..... $10 00

Spool for reloading only.. 2 00

THE EASTMAN DRY PLATE AND FILM CO.,

Rochester, N. Y.

15 Oxford Street, London.

Advertisement for the Kodak camera with a new slogan.

each "snapshot," the film would be advanced until the final number appeared in the red window. A roll of film could be of any size, but Eastman packed enough film in the camera to take one hundred circular photos, each two and a half inches in diameter.

YOU PRESS THE BUTTON . . .

In June 1888 Eastman introduced his revolutionary new camera. For twenty-five dollars anyone could buy one, already loaded with film. When the film was used up, the entire camera was shipped back to Rochester for processing. The film was removed, the photos developed and printed, the camera reloaded and shipped back ready to take another one hundred pictures—all for ten dollars. To help users keep track of their shots, a small memo book was packed with each camera. "You press the button, we do the rest" became the motto of the new Eastman Company.

George needed a brand name for his novel camera. Finding a name for a product can be even more difficult than inventing one. Eastman knew that no matter what he called his camera, it would have to have the letter K in the name. He just happened to like that letter. The *Kodak*, easy to use, easy to spell, was introduced in 1888. It swept the country. Here was a small, lightweight camera that even a child could operate. No messy chemicals to mix. No heavy glass plates to lug around.

But the Kodak wasn't good enough. The box camera was too expensive. For fifteen dollars, a man could buy a new suit. For fifty dollars down, a five-room cottage, featuring "four closets and a full cellar" and only eighteen miles from New York City, could be purchased (total price fifteen hundred dollars).

And the paper-based film nagged at Eastman's sense of quality. So in 1889 he marketed the first transparent roll film, made from cellulose. Then came smaller and smaller cameras: the pocket Kodak in 1895 and the folding pocket Kodak in 1897 (considered the ancestor of all modern roll cameras).

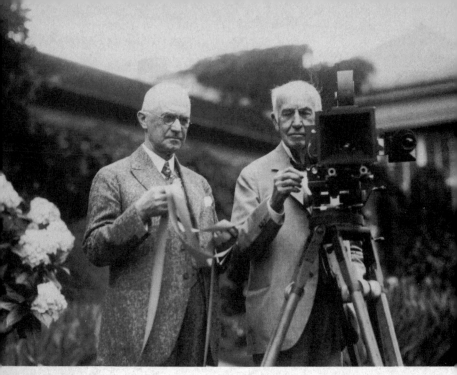

Thomas Edison (right) had George Eastman (left) to thank for developing the film (no pun intended) that made motion pictures possible.

THE BIRTH OF THE BROWNIE

By 1896 Eastman Kodak (the new company name) had sold more than 100,000 cameras and Eastman was standing at the brink of fulfilling his dream. In 1900 he introduced the camera that would revolutionize picture taking: the Brownie. Priced at just one dollar, the Brownie made picture taking so easy even a child could do it. Just aim and push the button. Rewind the film and send it back for processing. And cheap? A six-exposure roll of film sold for just fifteen cents.

The Brownie was an overnight sensation. Millions of shutter-bugs who never would have dreamed of taking their own pictures bought the new cameras. Kodak became the film powerhouse. Some 80 to 90 percent of all roll film sold in the world in 1901 was made by Kodak.

In 1889 Kodak patented a flexible transparent base for roll film. W. K. L. Dickson, an assistant to Thomas Edison, ordered a Brownie. What he really wanted was the film. Kodak's film was strong and flexible enough to be made in two-hundred-foot long rolls. When Edison saw the film, he knew he had found the material he needed to design a motion picture camera, the Kinograph, and the Kinetoscope, the forerunner of the theatrical film projector (it was actually a peep show—viewers had to peek into small cutouts to see the action).

EPILOGUE

Ironically, history has recently repeated itself. In 1987 Kodak introduced a new line of cameras already preloaded with film: the Kodak Fling camera, made to be used one time. When finished, the camera must be sent back—with its film—for processing. The film is removed, processed, and printed.

Alas, Kodak had turned back the clock almost a hundred years and reinvented the original Kodak.

What would George have said?

5

....

The Blendor That Conquered Polio

*At the bottom of this jar there's a little revolving band-shaped
thing like a propeller [turning at] 12,000 revolutions a minute.*

—FRED WARING, concocting a
banana-chocolate milk shake on
WEAF Radio, New York, 1938

CUISINART IS TO food processor as _____ is to blender.
The answer, of course, is Waring. The Waring Blendor. This
chrome-based mixer with the glass top, the predecessor of the modern food processor and the staple of bars and restaurants, found its
home in millions of American kitchens following World War II.

But guess who it was named after. Time's up. If you guessed
Fred Waring, leader of the famous Pennsylvanians band and choral
group, pour yourself a pina colada. Fred Waring made a crusade out
of promoting "his" Blendor, as anyone who was around in the
prewar era will tell you. And the name Waring eventually became
synonymous with his Blendor.

The truth of the matter is that Fred, an engineering student at
Penn State and a self-proclaimed gadgeteer, did not invent the
Waring Blendor (though he did put the O in the word Blendor). But
Waring was wise enough to recognize a good idea, and embraced
and developed it.

MALTED MILK SHAKES

The concept of swirling beverages around in an enclosed vessel
was the brainchild of Stephen J. Poplawski of Racine, Wisc. For

more than fifty years, dating back to 1915, Poplawski was fixated on the idea of creating various kinds of beverage mixers.

He applied for a patent in 1922 for his first mixer "having an agitating element mounted in the bottom of a cup and a driving motor mounted in the base." Recognize this as the nuts and bolts of the blender idea? Poplawski did, but only to a limited extent. He received his patent but envisioned his mixer for the concoction of soda fountain drinks such as malted milk shakes. "I did not think of using the mixer [based on] this patent, for the maceration of fruits and vegetables." Obviously no one else at the Arnold Electric Company in Racine, where he worked, did either. All eyes were firmly entrenched on capturing a larger share of the soda fountain business.*

But inventor Frederick J. Osius saw other possibilities. Back in the summer of 1936, Osius produced a contraption he called a "disintegrating mixer for producing fluent substances."[1] The crude blender needed development.† And like thousands of inventors before him, he had a novel idea but lacked the money to tinker with it.

Where to go? One of Osius's investors was the brother-in-law of Fred Waring's publicist. Why not hit the famous bandleader for money? It just so happened that at the time Fred Waring was playing a radio gig at the Vanderbilt Theater in Manhattan. Here was a golden opportunity. Faking an appointment, Osius talked his way into Waring's dressing room backstage. Having a reputation for a fondness for gadgets himself, Waring listened to Osius's tale of woe.

Ever the smooth talker, Osius produced a prototype of the mixer and attempted to sell and demonstrate at the same time. The machine failed to function. But Osius's sales pitch didn't fail. Waring was convinced by this eccentric stranger wearing striped pants and a bright lemon-yellow tie to put up the money for an invention he believed would change the way people mixed their drinks. A born

---

* Poplawski later went on to develop another mixer that would be marketed by the John Oster Manufacturing Company. Years later, in a patent fight, the courts would rule that Waring could not claim that Poplawski deserved to be known as the inventor of the blender because of its limited use aimed at mixing fountain drinks.

† On January 3, 1933, Osius received his first patent for a "drink mixer."

banana lover, Waring would later say, "I had been wanting a machine to make absolutely velvet-like, sweetest-possible banana milk shakes all my life."

PERFECTING THE DESIGN

Unfortunately, the prototype was not ready for production. Banana milk shakes would have to wait. Twenty-five thousand dollars and six months later, the blender still wasn't working properly. It leaked too much. Why? Look at the bottom of the glass container on any blender and you can see the challenge for yourself. Since the drive shaft goes through the glass, a major problem is keeping the liquid from leaking past the spinning blades and into the motor. Osius couldn't find the right leak-proof seals and couplings for the base of the container. So the blender leaked too much.

Waring brought in his colleague Ed Lee, who solved the mechanical problems and fixed the leaks. Next came a marketing makeover. No ugly prototype would ever sell; a new look was needed. Peter Muller-Munk, a German designer, gave the blender its famous polished art deco look of chrome and glass. And the Miracle Mixer, as it was called, was born. Making its world debut at the National Restaurant Show in Chicago in 1937, the blender was marketed as a gadget for making frozen daiquiris and other mixed drinks. Its price of $29.75 was a small fortune, equal to about a week's pay at Depression-era salaries.

Christened the Waring Blendor, with an O, the device was hyped by Fred Waring at every chance. While touring the country with the Pennsylvanians, Fred used his dressing room to hawk his wares. The bandleader even had a theatrical trunk customized to include a complete bar for demonstrating the drink-mixing capabilities of the new Blendor, as well as what one reporter called a "temperance" bar in his New York City office. (Fred was a lifelong teetotaler.)

Grinding up just about everything he could get his hands on, Waring made bizarre concoctions backstage. Members of the band avoided his dressing room lest they be forced to sample the latest.

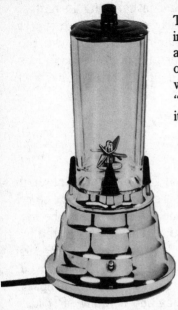

The original Waring Blendor, as seen in a 1938 Hammacher Schlemmer advertisement. In 1959 a U.S. Court of Appeals ruled that the Blendor was different than other food "mixers." Even so, the ad still called it the Waring Mixer.

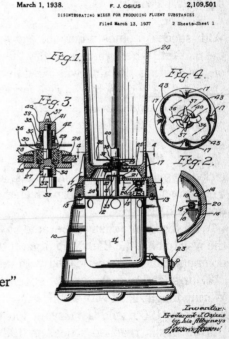

March 1, 1938.   F. J. OSIUS   2,109,501

DISINTEGRATING MIXER FOR PRODUCING FLUENT SUBSTANCES

Filed March 13, 1937   2 Sheets-Sheet 1

Patent for "disintegrating mixer" filed in 1937 by Fred Osius.

Not so fellow crooner Rudy Vallee. Intrigued by the food processor himself, Vallee became one of its greatest supporters.

The war depressed the Blendor market. With sales down, Waring sold the company to entrepreneur Hazard Reeves in 1947. Why keep the Blendor only in bars? reasoned Reeves. In a stroke of marketing genius, he opened up the home market. An immediate success, the one-millionth Waring Blendor was sold in 1954.

Soon the Blendor was finding uses other than making the perfect daiquiri. Hospitals were among the early customers, proving it unequaled for preparing baby food. Modified for laboratory use, the newly renamed Waring Aseptical Dispersal Blendor helped Jonas Salk combat polio when used to grind up materials needed to prepare cultures for the polio vaccine.

BLENDER WARS

With success comes imitation, and soon rival blenders saturated the market. Where there's rivalry there are lawsuits, and the story of the Waring Blendor was no different. At issue was whether Osius's idea represented a real innovation that was protected by law. The courts agreed it did. A United States Court of Appeals ruled in 1959 that "Osius was the first to achieve . . . an improved apparatus not only to mix liquids but also effective to accomplish rapid and thorough disintegration, mixing and aerating of solids, including pulpy and fibrous materials, and fluent substances to form a uniform and creamy blend in which the solid material is thoroughly emulsified."[2]

With that court ruling it became obvious to the competition that the way to beat the Blendor barons at their own game was to create competing models overloaded with new features: lots of buttons.

Four-speed and then eight-speed blenders appeared in 1964. Waring fought back with a solid-state model in 1965, a nine-button beauty in 1966, and a dizzying fourteen-speed model in 1972. Button mania reached a peak in 1989 when Waring's deceptive eight-button model highlighted a breathtaking dual-speed feature giving it a dazzling sixteen operating speeds. Who could believe that anyone could have lived with just the original dual-speed model?

6

....

From a Melted Candy Bar to Microwaves

One sometimes finds what one is not looking for.

> —SIR ALEXANDER FLEMING,
> quoted in *People's Almanac*,
> Vol. 2, 1978

NEXT TIME YOU nuke a bag of Orville Redenbacher's, you'll be repeating an experiment that heralded the dawning of the age of microwave cooking.

Almost fifty years ago, 1946 to be exact, one of the great minds in the history of electronics accidentally invented microwave popcorn.

Shortly after World War II, Percy L. Spencer, electronic genius and war hero, was touring one of his laboratories at the Raytheon Company. Spencer stopped in front of a magnetron, the power tube that drives a radar set. Suddenly he noticed that a candy bar in his pocket had begun to melt.

MELTS IN YOUR POCKET, NOT IN YOUR HANDS

Most of us would have written off the gooey mess to body heat. But not Spencer. Spencer never took anything for granted. During his thirty-nine years with Raytheon, he patented 120 inventions. When England was battered by German bombs in the 1940 Battle of Britain, Spencer turned his creative mind toward developing a better version of the British invention radar. His improved magnetron allowed radar tube production to be increased from seventeen per week to twenty-six hundred per day. His achievements earned him the Distinguished Service Medal, the U.S. Navy's highest honor for civilians.

So when this inquisitive, self-educated, and highly decorated engineer who never finished grammar school came face to face with a good mystery, he didn't merely wipe the melted chocolate off his hands and shrug off the incident. He took the logical next step. He sent out for popcorn. Holding the bag of unpopped kernels next to the magnetron, Spencer watched as the kernels exploded.

EXPLODING FOOD

The next morning Spencer brought in a tea kettle. He wanted to see what microwaves would do to raw eggs. After cutting a hole in the side of the kettle, Spencer placed an uncooked egg into the pot. Next he placed a magnetron beside the kettle and turned on the machine.

An unfortunate (cynical?) engineer poked his nose into the pot and was greeted by an explosion of yolk and white. The egg had been blown up by the steam pressure from within.* Spencer had created the first documented microwave mess—an experiment to be inadvertently repeated by countless thousands of microwave cooks. He had also shown that microwaves had the ability to cook foods quickly.

Legend has it that this demonstration was reproduced before unsuspecting members of Raytheon's board of directors who had trouble visualizing exactly what microwaves could do to food. The ensuing egg shower convinced the board of directors to invest in the "high frequency dielectric heating apparatus," patented in 1953.

* Thus the myth was born that microwaves cook food from the inside out. Microwaves cook food by shaking up its molecules rapidly. To say food cooks from the inside is not exactly true. Microwaves may cook the outer layer of a frozen steak but dissipate before ever entering the meat's middle. The result is a frozen center and a cooked surface.

The first microwave ovens were too big for the home. Tilting the scales at 750 pounds, this one found its home on ships and trains.

That demo and the fact that the military no longer needed ten thousand magnetron tubes per week for radar sets helped shape the future of microwaves. What better way to recover lost sales than to put in every American home a radar set disguised as a microwave oven?

But first the device needed a better name. Raytheon's marketing mavens felt few people would demand a high-frequency dielectric heating apparatus for their kitchens even if they could pronounce it. A contest followed to rename the apparatus. Seeing as how the oven owed its roots to radar, the winning entry suggested "Radar range." The words were later merged to Radarange. But no words could hide the woeful inadequacies of this first-generation oven.

CUTTING IT DOWN TO SIZE

Weighing 750 pounds and standing five and a half feet high, the Radarange required water—and plumbing—to keep its hefty innards cool. Hardly the compact unit that fits under today's kitchen cabinets. The early 1953 design—with its three-thousand-dollar price tag—was strictly for restaurants, railroads (the Japanese railroad system bought twenty-five hundred), and ocean liners. These customers would be Raytheon's prime market for two decades.

The microwave oven was no pleasure to cook with, either. Culinary experts noticed that meat refused to brown. French fries stayed white and limp. Who could eat this ugly-looking food? Chefs were driven to distraction. As chronicled in the *Wall Street Journal*, "the Irish cook of Charles Adams, Raytheon's chairman, who turned his kitchen into a proving ground, called the ovens 'black magic' and quit."

It would take decades before the consumer oven was perfected. The Tappan Company took an interest in the project and helped Raytheon engineers shrink the size of the magnetron. A smaller power unit meant the hideous plumbing could be done away with and air cooling fans could take over.

RADAR IN A BOX

Then someone had the brilliant idea that perhaps the magnetron should not be pointed directly at the food but rather out of sight. That's it. Put the food in a box, put the microwave source at the back, and lead the microwaves into the box via a pipe. Now we could truly call it an oven.

And that's what happened. In 1955 Tappan introduced the first consumer microwave oven. Did you have one? Hardly anyone did. It was still too big and costly. Then came 1964 and a breakthrough. From the country that had a reputation for making "transistorized" (read: small) products out of everything, Japan, came an improved electron tube. Smaller and simpler than the old magnetron, it put Raytheon on track to placing a microwave under everyone's kitchen cabinet.

Needing a consumer-oriented vehicle to sell its new ovens, Raytheon bought up Amana Refrigeration, Inc., in 1965 and put out its first affordable ($495), compact, and practical microwave oven in 1967.

**5-pound rib roast.
Medium rare.
30 minutes.**

The first counter-top microwave ovens came only after years of shrinking down the original models.

IS IT SAFE?

The specter of little microwaves leaking out of the oven scared a lot of people. Their worst fears were realized in 1968 when a test of microwave ovens at Walter Reed Hospital found microwaves did indeed leak out. Federal standards set in 1971 solved that problem.

Today more homes have microwave ovens than dishwashers. And we owe it all to an inquisitive man with a melted candy bar in his pocket and egg on his face.

7

....

Fax: The Priest and the Pendulum

The owner of a circus wired his man in India saying: "I need macaque monkeys; please send two." A few days later he received a reply: "I am sending you 1500 macaques, 500 more will come next month."

> —R. CHAMPEIX, *Savants Méconnus,*
> *Inventions Oubliées*, an example of
> errors created by telegraphy

QUICK. WHICH IS older? The telephone, the radio, or the fax machine? If you guessed anything but the fax* machine, you're wrong! Hard as it is to believe, the fax (facsimile) machine is the oldest, even older than the telephone. Messages were being faxed more than thirty years before Bell's telephone! Patents for the first fax machine date back nearly 150 years, to 1843. Of course the first fax machines worked a lot differently than they do today. But they still produced magnificent reproductions.

The first commercial fax system was invented by an Italian priest, Giovanni Caselli, who helped establish a fax line between Paris and Lyons for five years, 1865–70. Giovanni Caselli was considered by his neighbors as "some kind of a nut" (this moniker is worn proudly by many inventors). With scientific "junk" strewn about the furniture, the Italian priest's small home looked more like a mad scientist's workshop than the residence of a man of the cloth.

Born in 1815 in Siena, Caselli had simultaneously pursued both theological and scientific studies. Typical of the kind of genius who

---

* The terms *fax* or *telefax,* short for facsimile, became accepted pieces of technical jargon only in 1980.

couldn't leave well enough alone, Caselli found himself immersed in politics, and before long his political leanings forced him into exile in Florence in 1849.

The telegraph had opened the world to the convenience of sending messages by code. And now inventors turned their attention to a bigger problem: Sending pictures by wire was the talk of the

World's First Fax Machine. Patented in the United States in 1863—before the telephone—this fax machine, called the pantelegraph, was sending faxes from 1865 to 1870 in France. The original was a manuscript written directly on a metal plate with insulating ink and placed on a small platform shown on the right side of the machine. The first prototype fax machine was patented in 1843.

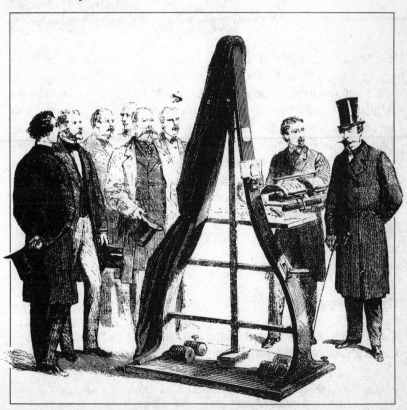

Giovanni Caselli. Italian priest who patented the first practical fax machine operating on a telegraph line between Paris and Lyons from 1866 to 1870. "The most delicate traces of writing and drawing may . . . be reproduced immediately," wrote Professor Auguste de La Rive in 1858.

electrical community of the day. And Caselli could not resist jumping in with an idea of his own: the "pantelegraph."*

Caselli resurrected an idea first thought up by Alexander Bain of Scotland in 1840. Shortly after the invention of the telegraph, Bain invented a crude method of sending pictures over telegraph wire. Bain was an expert clockmaker, so his fax machine naturally involved swinging pendulums connected at each end of a telegraph line. For a character to be sent by wire, it first had to be made out of metal. At the sending side, a pendulum tipped with a metal stylus swung back and forth over the character. On the other end of the line, a similar pendulum swung back and forth across chemically treated paper.

FAXING BY PENDULUM

Each time the sending pendulum struck the metal, it completed an electrical circuit and sent an electric pulse through the telegraph

---

* From the Greek *pan*, meaning "all," and the word telegraph: a telegraph that could transmit all types of documents, words and pictures; it would "write everything."

wire. Arriving at the swinging pendulum on the receiving end, the electricity discolored the chemically treated paper, leaving a trace on the paper each time contact was made. Clocks on both ends advanced the metal characters and the paper a fraction of an inch in unison, or so went the theory of operation. But in reality, it was impossible to synchronize both pendulums and clocks. The machine was never developed (although patents for the facsimile machine date back to the 1840s), but the kernel of an idea was planted that would later lead to a commercially successful fax machine.

Instead of using a flat plate with an awkward pendulum, English inventor Frederick Bakewell hit upon the idea of a rotating drum. The message was written on the drum with a nonconducting "ink." The metal sheet was wrapped around a revolving cylinder. A metal stylus was set on the cylinder and a screw mechanism kept the stylus moving along at a uniform rate.

As the stylus glided along the bare metal, it emitted an electric current. When it touched the insulating ink, the current was shut off. At the receiving end a sheet of chemically treated paper was rotating on a similar drum. The electrical impulse discolored the paper and caused a reproduction of the original. The electric current produced a blue trace on the paper; no current left a white trace. The fax appeared as white on blue.

Unfortunately, this design also suffered from a finicky synchronization problem, and while Bakewell patented his fax in 1848, the device never was commercialized.

Enter Giovanni Caselli. Hiding in Florence from his political enemies, Caselli had found a comfortable teaching position at the university. Not content merely to teach, Caselli tried his hand at publishing. He founded a crude scientific journal called *La Ricreazione*, which allowed his novel ideas to fall under the gaze of Florence's elite and educated.

Caselli resurrected Bain's clock-driven pendulum idea but, lacking the skills to make a working prototype, he set off to Paris to find one of the world's leading builders of scientific equipment, Gustav Froment. Caselli and Froment labored for seven long years, 1857–64, refining and improving the fax machine. The product had to be

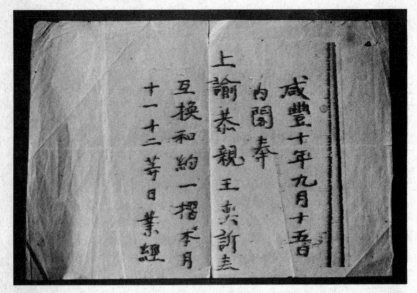

First Fax Cover Sheet? One of the earliest fax (pantelegraph) cover sheets, with a message in Chinese:

September 15, the 10th year [1861] of the Xian Feng Reign [1851–61]
  To: The Cabinet Majesty Emperor Xian Feng
  From: [?]
  Re: Copy of treaties documented on the 11th and the 12th day of this month.

easy to use and reliable. Caselli had to put up with the snickering of the scientific community, which said it couldn't be done.

Finally, in 1863, Caselli could laugh at the skeptics. He had finally triumphed. The receiver of a U.S. patent for his "telegraphic apparatus," Caselli made major improvements to the fax design:

☞ He installed a separate clock in the transmitting end that synchronized the pendulums on each end perfectly.
☞ His fax did not require the original to be scratched in metal or written in special ink. Ordinary ink sufficed.
☞ Originals could be reproduced in the same size or reduced.
☞ Different messages could simultaneously be transmitted through a single wire.

Also among the improvements was Caselli's very high quality recording paper. It was soaked in potassium cyanide, which changed color each time electricity passed through it. The result was a fax machine standing more than six feet tall, composed of swinging pendulums, batteries, and wires. But this ugly duckling produced faxes of outstanding quality. Great quality, coupled to a system that didn't require a lot of maintenance, meant the fax machine was now a marketable product.

THE FIRST FAX LINE

The government of Emperor Napoleon III liked what it saw and decided to embrace the pantelegraph as its own. The French legislature passed a law calling for service between Paris and Lyons. In 1861 the government authorized tests of a fax system using telegraph lines between Paris and Lille and Paris and Marseilles.

By 1863 a Paris-Lyons line was tested with great success. Transmitting at fifteen words per minute, the fax could send forty telegrams of twenty words each hour. In 1865 the French government decided to take the system public. Officially inaugurated on May 16, 1865, the pantelegraph was set up on the existing Paris-Lyons telegraph line. The French must have been exhilarated because two years later, the Marseilles leg was added. By 1867 four Caselli machines serviced the Paris-Lyons lines. Service was so improved that by 1868, some 110 telegrams could be sent each hour. While ornate and flowery pictures were occasionally faxed, the fax line generally carried stock market information between 1866 and 1870.

The fax caught the eye of the press (perhaps because they could see its benefits to their industry). Newspapers and magazines praised the new invention for its public service. "Who could do better?" wrote the journalists. But the fax machine did not mushroom into a commercial hit. Why? No one can say for sure. Certainly it was a technical success. But technological breakthroughs are no guarantee of consumer demand—or else today we'd all be communicating by picture phone. Perhaps the fax didn't fill a real need, or maybe it was just a novelty. Maybe competition from the telegraph proved too great.

Interrupted by the 1870 war and the siege of Paris, the fax system was not reinstalled. The era of the mechanical fax machines with their pendulums and clockworks was coming to an end. The turn of the century saw the light, literally, of the modern fax. Machines that used light beams to read and write would replace the swinging pendulums.

PICTURES BY LIGHT

In 1873 Willoughby Smith discovered that the dark gray toxic mineral called selenium would give off an electric current when light shone on it. Why not make a mosaic of selenium cells, project a picture on the mosaic, and send an image by wire? That's what G. R. Carey of Boston did in 1875. An image was focused on the mosaic. Each selenium "photocell" gave off an amount of electricity in proportion to the intensity of the light falling on it. And sure enough, when a corresponding mosaic made of shutters was constructed on the receiving end, the first crude light-derived image was sent by wire. (Was this really the first television image? The difference between fax and television is the speed at which the images are sent. TV sends thirty frames per second. Fax takes a few seconds per frame.)

Noah Amstutz of Cleveland showed in 1891 that a picture could be reproduced using photocells made of selenium. The transmitter used a single selenium cell and the receiver reproduced pictures by means of dots of various half-tones.

In 1902 physicist Arthur Korn of Germany improved the idea dramatically. Korn retrieved Bakewell's rotating drum idea. He constructed his cylinder out of glass, positioned a selenium cell inside, and wrapped the intended fax around the cylinder. By shining a light through the picture and onto the selenium, Korn's machine accurately converted the picture into electrical signals. On the other end, the signal was converted back to light and was shone on photographic film. Voilà. A permanent copy.

Korn's system was an overwhelming success, especially for the newspaper business, which was rapidly growing in circulation. Now instead of hand-drawn illustrations, readers could see real photographs in their morning papers. The fax machine was rapidly becoming indispensable to the press.

Realizing the machine's value in the competitive newspaper business, the French newspaper *L'Illustration* bought out Korn's rights to the machine in 1906 and monopolized its use in France.

THE FAX GOES PORTABLE

The drawback to photocell fax was that the pictures had to be on a special set of wires. Couldn't it be possible to make use of the tens of thousands of miles of telegraph wires already in use? That's what Édouard Belin thought. Belin invented his own "telegraphiscope" in France just slightly behind Korn. Not being able to crack Korn's monopoly, he dropped the idea. In the best spirit of not getting mad but getting even, Belin eventually spirited away the newspaper business from Korn in 1913 with the world's first portable machine called the Belinograph, known to its friends as the Belino.

Simpler than Korn's machine and smaller than a typewriter, the Belino had unmatched advantage: It could be hooked up to common telephone lines. Now any reporter in the field could fax back photographs anywhere in the world.

World War I was a critical test for Belin's machine. The Belino made history by sending back the world's first remote fax photo/news report in 1914.

PHOTOS BY PUNCHED TAPE

H. G. Bartholomew and M. L. D. MacFarlane, two British inventors, found a different way of sending faxes by phone: converting and storing the images on punched tape. The tape was then fed through standard telegraph transmitters. On the other end, a similar device decoded and printed the images. Hence the newspaper wire service was born. Western Union, by 1924, was regularly sending photographs to newspapers by wire facsimile.

At one point following World War II, newspapers toyed with the idea of faxing the newspaper right to your home. But the mushrooming television industry put that idea to rest. (Newspapers have resurrected that idea and now fax abridged editions to selected subscribers. Timing is everything, isn't it?)

The problem with early fax machines was that an owner had to buy the sending and receiving machines from the same maker. Competing brands couldn't talk to one another. All that changed in 1974, when the first international fax standard was invented by the United Nations. Called Group 1, it allowed fax machines to exchange messages at a rate of six minutes per page.

TODAY'S MACHINES

In 1980, modern office fax machines came into being with the advent of Group 3—a fax standard that allows digital signals to be sent over regular telephone lines in one minute or less. If you own a fax machine, it uses Group 3. The pictures or text are converted to series of computerized bits—zeros and ones if you will—and sent via standard telephone lines. On the other end, the fax machine decodes the bits and reconstructs the image.

But just around the (far) corner is Group 4. Group 4—if it becomes accepted—will allow for higher sending speeds, fewer errors, and the ability to send color photos. However, Group 4 now suffers the ills plaguing early fax machines: no way (yet) to send it over regular phone lines, and competing machines are incompatible. Stay tuned.

8

....

How Not *to Invent the Telephone,*
or
Why Western Union Didn't Become Ma Bell

As to Bell's talking telegraph, it only creates interest in
scientific circles . . . its commercial values will be limited . . .

> —ELISHA GRAY, telephone
> inventor runner-up, June 1876

OF ALL THE "happy accidents" in the history of science, none has been more ballyhooed or immortalized in print and film than the invention of the telephone. Yet despite the amount of ink spilled and celluloid burned, no moment of discovery in the history of the United States has been more misconstrued than Alexander Graham Bell's infamous talking machine (my apologies to Don Ameche).

In fact, one experienced inventor of Bell's era charged that he, not Bell, was the inventor of the telephone. Elisha Gray took Bell to court in the 1870s, suing for the patent rights to the talking machine. Of course Gray did not win the patent fight; Bell is credited with the invention. But Gray's claim can hardly be ignored. Sketches of his telephone, produced weeks before Bell's patent, are remarkably similar. (See page 72.) And history might have recorded Gray as the inventor if Bell had not beaten him to the patent office by a few hours.

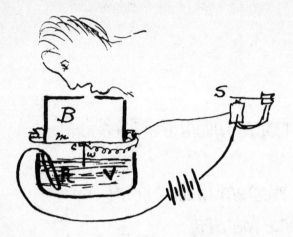

Almost Identical. Sketches of telephone transmitters made by Elisha Gray (bottom) and Alexander Graham Bell (top) show two inventors with the same idea. Gray sketched his telephone on February 11, 1876, two months after he envisioned it. Bell's sketch comes from his notebook, dated March 9, 1876. Bell received the patent for the telephone, beating Gray's patent application to the patent office by two hours. Even though Gray could have submitted his idea earlier, he believed the phone to be nothing more than a toy.

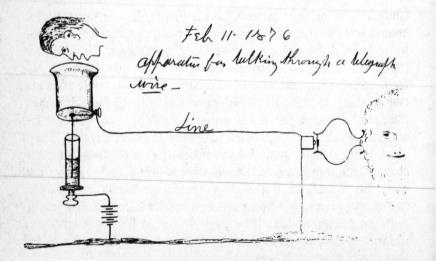

THE TELEPHONE'S ONLY A TOY

Ironically, Gray could have easily won the race if he had believed that the telephone was a useful device. He understood the technology needed to make it a success but didn't follow up. He couldn't see how a device that carried speech could be anything more than a toy. Until it was too late. Therefore, the race for the telephone is the story of how a professional school teacher, working in his spare time, beat out a highly successful professional inventor who had already been awarded his first patent.

The story is one of the most fascinating and illustrative of the contrast in style between two different kinds of men: one steered by the promise of fame and fortune, the other a purist believing knowledge to be its own reward. One who rigidly followed a narrow path of work, the other more able to adapt to new ideas.

FINDING A BETTER TELEGRAPH

Neither man had the telephone as his goal. The mother of this invention was the need to develop a better telegraph system. Introduced in 1844, the telegraph was such a huge success that tens of thousands of miles of telegraph wire had been strung across the country. But the carrying capacity of each wire was limited to one message at a time. The "dits" and "dahs" of Morse code were transmitted by opening and closing a switch. The Morse code wiring, being less sophisticated than today's light switch, could only send or receive one message per wire. So if you wanted to send more messages down the line, you had to install more wires to do it. One for each transmitter. One for each receiver. The limitations of the system implied that to install the telegraph in any large city required a rat's nest of wires that would choke the city in a sea of copper. More importantly to businessmen, it would be very expensive.

So the industrial giants of the day were looking for technology to come to the rescue, looking for someone to invent a way of carrying more messages per line, a system called multiplexing. A partial solution arrived in 1872 when Western Union installed the Stearns duplex, a way of sending a signal from each end of a wire at the same time. But true fame and fortune awaited the inventor who could find a way of sending four or more messages on one wire.

Enter our main characters: Elisha Gray, a professional inventor from Chicago, and Alexander Graham Bell, an elocution teacher and amateur tinkerer newly arrived in Boston.

Arriving in Boston in the spring of 1871, the twenty-four-year-old Bell set out to teach at his father's school for the deaf. Alexander Melville, the elder Bell, was a professor of speech and author of many textbooks on speaking correctly. He specialized in working with the hearing impaired and invented a method of teaching the deaf to speak known as "visible speech."

Alexander Graham Bell was a born inventor. He and his brother spent the first few years of their lives tinkering with models of the human voice. They once fashioned their own version of a set of vocal cords into a human skull. It's even reported that the device could imitate the word mama so well that a neighbor came running to see what was the matter with the "baby."[1]

BELL AND HIS TUNING FORKS

Many great inventors and scientists have their curiosity piqued by new toys. For Einstein, it was a compass given to him as a young boy that is credited with turning his thoughts toward science. For Bell, it was the tuning fork. As a youth, Bell had experimented with notes produced by the vibrating tongs, and the tuning fork would later play a key role in the invention of the telephone.

All the elements began to jell in 1872. Bell read in the newspapers about Western Union's duplex telegraph system and realized the riches that awaited the inventor of a better system. Convinced that he could invent a system that could transmit a dozen or more tones per wire, Bell turned to the tuning fork for the answer. He had studied the work of the German physicist Hermann von Helmholtz, who had invented an electric tuning fork. By hooking up a tuning fork to an electromagnet and plucking the fork, Helmholtz had found that an electric current vibrating at the same frequency as the fork could be sent through the wire.

Why shouldn't it be possible, reasoned Bell, to hook up lots of tuning forks to one wire and have each one carry a separate signal? Tuning forks vibrate at unique pitches—producing such diverse notes as C and G for example. Perhaps the fluctuating electrical signals from one fork would not interfere with the fluctuating signals of the other fork on the wire. All that was needed was a way of filtering out one signal from another on the wire. That was easy, reasoned Bell. A set of receiving reeds attached to the end of the wire would do the job.

Each reed, tuned to the exact frequency of each sending fork, would respond only to its own frequency. So the number of signals that could be sent over one wire was limited only to the number of notes on the diatonic scale. Voilà: the "harmonic telegraph."

Elegant and simple. And exactly the same conclusion reached independently by Elisha Gray.

ELISHA GRAY

Born in Barnesville, Ohio, in 1835, Gray shared an early interest in telegraphy. But Gray's interest—as opposed to Bell's—was totally professional. At the age of thirty-two, Gray received his first patent for a telegraph relay. Realizing the value of his work, Gray set up his own business, the Western Electric Manufacturing Company, to produce telegraphic instruments. The giant telegraph company, Western Union, kept a close watch on Western Electric. Hoping to gain control of the field, Western Union bought a one-third interest in the business in 1872.

More importantly, through his association with Western Union, Gray realized the pot of gold that awaited the inventor of the multiplex telegraph system.

THE BATHTUB EXPERIMENT

In 1874 Gray made a major breakthrough. In what later came to be called an improved version of his "bathtub experiment,"* Gray showed it was possible to send different musical notes over wire and have them be reproduced on the other end. In effect, he had invented an early version of the electric organ. He realized that through "musical telegraphy" he could harness that idea to go one of two ways: Use each tone to carry many signals over one wire, or devise a method for sending speech.

The implications of these inventions were not lost on the public, which embraced the idea of the telephone. Demonstrations of his transmitters and receivers in Washington, Boston, and New York caused the *New York Times* to quote one Western Union official as saying that soon telegraph operators "will transmit the sound of their own voices over the wire, and talk with one another instead of telegraphing."

THE TELEPHONE: NO PRACTICAL VALUE

Private industry was not so enthusiastic. Believing speech transmission to be a waste of time, the top technical journal of the industry, *The Telegrapher*, put down the idea, claiming it was not new and the telephone had "no direct practical application."

Even Gray's colleagues were not impressed. His patent attorney told him the telephone was at that time a scientific curiosity.

Under the weight of that criticism and with the advice of his

---

* In early 1874, Gray found his nephew playing with some electrical equipment in his metal bathtub. He was holding the live end of an electrical "shock" coil in one hand while running his other hand over the zinc-lined bathtub. Gray noticed that a sound came from underneath the hand. Repeating the experiment, Gray found that the sound emanating from the tubbed hand carried the same pitch as the vibrating apparatus that powered the shock coil. This set him thinking that "vibratory" currents could be transmitted through wires. To what practical use this discovery could be applied would take months to crystallize in his mind.

business partners, Gray gave up the idea of the telephone as a money-making enterprise and concentrated on the multiple telegraph. He had reached the conclusion that speech could be transmitted by wire months before Bell had, but he now set the idea aside. He was convinced that businessmen could do more useful work by sending messages than by talking.* It would be a decision he would later regret.

TIN CAN TELEPHONE

Ironically, in the fall of 1875 Gray would find the missing ingredient in his half-hearted effort to invent the telephone. He observed two boys playing with a homemade telephone consisting of two tin cans and a string. A boy talking into one tin can sent vibrations in the metal that could be carried by the string to the other can. This was the answer he had been looking for. He already had a very good receiver. What he lacked was a suitable voice transmitter. The tin cans were the spark of genius. He sketched a voice transmitter using a tin can–like voice chamber hooked to a wire. The wire was attached to an electrically conductive liquid whose resistance varied with the vibrations of the can. Such a setup could change speech into electrical signals that could be transmitted by wire. Gray had solved the problem of transmitting speech. But his skepticism about the usefulness of the telephone held him back from patenting the idea.

He waited over two months to put the idea on paper—until February 11, 1876. When he finally decided to protect his idea by patenting it, Gray found his drawing arrived at the patent office two hours behind Bell's patent application for the telephone on February 14, 1876.

* Perhaps history will bear him out. With the meteoric rise in the use of fax machines, Gray may have foreseen how much more useful sending printed material over a phone line is than taking one's chances with the vagaries of speech.

THE BREAKTHROUGH

Bell had gone in the opposite direction. Although pressured by his financial backers to give up the telephone idea, he pressed ahead with it. By mid-1875, he gave up his work on the telegraph, proclaiming "the day is coming when telegraph wire will be laid on to houses just like water and gas—and friends converse with each other without leaving home." Bell would stand alone in his conviction.

Coming to that conclusion required a bit of luck. Had the harmonic telegraph been a well-behaved child, the telephone might never have been invented. As it was, it required a lot of tinkering. The break Bell had been waiting for occurred on an unseasonably hot afternoon, June 2, 1875. Here is what happened.

Working out of their small shop on the top floor of the Charles Williams's telegraph shop at 109 Court Street in Boston, Bell and Thomas A. Watson, his brilliant young mechanic, were laboring to send musical notes on their harmonic telegraph. Watson built a set of transmitters and receivers for their experiments. Each device consisted of an iron bar magnet set into a spool of wire. A tuned metallic reed—acting as a tuning fork—was held tightly in place a fraction of an inch above the coil by two screws (see illustration). Each transmitter's reed was tuned to a different note.

Bell placed himself in one room. Watson was stationed in another room at the other end of the building. The only connection between the two rooms was a set of wires running between the instruments. The halls and walls acted as sound buffers so as not to obscure the notes that were anxiously being awaited from Watson's device. Despite his best efforts, Bell could not get the harmonic telegraph to work correctly. The "mocking fiend inhabiting that demonic telegraph apparatus"—as Watson called the bugs in the system—prevented the notes from coming through clearly.

Bell found that the messages got mixed up because the sending and receiving reeds were out of tune. Their pitches were not exactly matched. So Bell set out to tune the vibrating steel reeds.

Professor Bell's vibrating reed—used for a receiver.

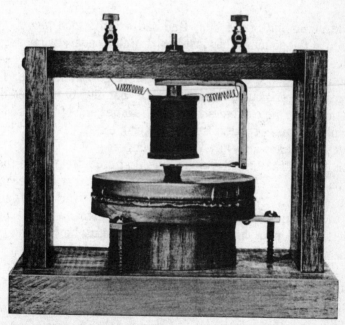

Alexander Graham Bell's first telephone.

"Bell always had to do this tuning himself," reflected Watson in recalling that evening, "as my sense of pitch and knowledge of music were quite lacking." Bell placed each receiver against his ear while the corresponding transmitter in a distant room was sending intermittent electricity through the magnet of the receiver. Watson described what happened on that hot afternoon of June 2, 1875:

> I had charge of the transmitters as usual, setting them squealing one after the other while Bell was retuning the receiver[s]. One of the transmitter[s] I was attending to stopped vibrating and I plucked it to start it again. It didn't start and I kept on plucking it, when I heard a shout from Bell in the next room, and then out he came with a rush, demanding, "What did you do then? Don't change anything. Let me see!"

SERENDIPITY STEPS IN

On the other end of the line, Bell had heard a loud *twang*. Astounded at the sound he has just heard, Bell wondered how such a tone could have come from a telegraph set. In plucking the reed, Watson had induced an electrical current to flow in the wire. The undulating electric current mimicked the vibration of the reed diaphragm. When the current arrived at the other end of the wire and encountered Bell's reed, it set it twanging exactly as it had been instructed (unknowingly) by Watson's finger.

This was a true breakthrough. The first crude telephone message had been sent. It is perhaps at that moment that Bell realized how close he was to transmitting speech by wire. Watson was sure that "the speaking telephone was born at that moment. That experiment showed him that the complex apparatus he had thought would be needed to accomplish that long dreamed result was not at all necessary, for here was an extremely simple mechanism operating in a perfectly obvious way that could do it perfectly." If a vibrating reed can be heard over a telegraph, why not human speech?

The idea that speech could be carried by wire had been surging back and forth in Bell's head for two or three years.* One evening years before, while resting between struggles with the telegraph apparatus, Bell told Watson, "If I could make a current of electricity vary intensity, precisely as the air varies in intensity during the production of a sound, I should be able to transmit speech telegraphically." He then sketched an instrument that he thought would accomplish this. Watson, ever the tinkerer, looked at the sketch and decided the device would be too costly to make and would probably not work. Besides, said Watson, the financial backers were only interested in perfecting the harmonic telegraph, not the telephone. Complete the harmonic telegraph, said Watson, and you "would have money and leisure enough to build air castles like the telephone."

The events of June 2 changed that thinking. The accident had presented a unique opportunity; Bell could no longer ignore its knock. He would give up on the telegraph and concentrate on the telephone.

TURNING TOWARD THE TELEPHONE

The two inventors were flushed with excitement. That night, Bell sketched for Watson an apparatus to be made the next day in hopes of actually transmitting speech. Following his instructions, Watson modified one of the devices by fitting a sound-gathering mouthpiece over it. To ensure quiet for the upcoming experiment (Watson's sense of hearing was acute), he ran the world's first telephone line from one of the rooms in the attic down two flights of steps to the third floor where Williams's main shop was, ending it near his workbench at the back of the building.

Bell assumed his position at the transmitter and shouted into the

---

* While experimenting at the Massachusetts Institute of Technology (MIT) a few years back, Bell had seen a model of an object called a "telephone" by German physicist Johann Philip Reis in 1861. Reis's telephone receiver was little more than wire wound around a knitting needle attached to a wooden box and incapable of delivering human speech. But Bell made note of the device and it may have set him to thinking about making his own telephone. Some historians, especially those in Germany, credit Reis as the father of the telephone.

makeshift microphone. Holding the crude device to his ear, Watson heard noise coming from his transmitter but couldn't make out any words. Over the next few days they improved the system. By July 1 they could "talk back and forth" and hear each other's "vocal sounds" but still not hear any intelligible speech. But improvements continued.

THE RACE TO THE PATENT OFFICE

On February 14, 1876, Bell submitted one of his designs to the patent office. The transmitter was an electro-mechanical affair, composed of magnets and wires. But Bell was not happy with the quality of the transmitter. It needed improvement.* On March 9, Bell sketched an improved liquid transmitter.

The next day finds the inventors on the top floor of a boarding house at 5 Exeter Place in Boston. One room serves as Bell's sleeping quarters. Another is his laboratory. A new liquid transmitter sits on a table in front of Bell. It is shaped like a cone, and the liquid in it is acid, highly corrosive acid. Just as they had done before, each assumes his separate position in widely spaced rooms. Unlike the countless times before, in which he speaks into the mouthpiece hoping Watson will understand his words, this time Bell yells, "Mr. Watson—Come here—I want to see you." Watson hears the words loud and clear, and the first message sent by telephone is immortalized in history.†

In an ironic twist of fate, Elisha Gray had independently designed his own "liquid transmitter." On February 11, 1876, just over two months after he envisaged the idea, Gray sketched a liquid transmitter literally identical to one sketched by Bell on March 9. (See

---

* Bell had such trouble with his transmitter that he eventually settled on a transmitter (mouthpiece) designed by Thomas Edison.

† Watson would later apologize for such uninspiring first words. Explaining that the unexpected occasion had not been arranged or rehearsed, Watson said that "if Bell had realized that he was about to make a bit of history, he would have been prepared with a more resounding and interesting sentence." Just a few years before, during the sending of the first telegraphic message, Samuel Morse had tapped out "What hath God wrought," which Watson had called an example of a "noble" first message.

sketches by Bell and Gray, page 72.) Had Gray believed in his own device, he could have drawn and submitted his sketch to the patent office weeks before the arrival of Bell's patent application. Instead he waited. And Bell's patent application arrived at the patent office two hours before Gray's. Gray declined to contest the patent on the advice of his patent attorney and financial backers that the telephone was merely a toy.

STILL NO INTEREST IN THE TELEPHONE

Even after winning his patent Bell could not convince big business, or Elisha Gray, of the telephone's usefulness. The 1876 Centennial Exhibition in Philadelphia featured Bell giving a masterly demonstration of his new telephone. He was singled out for special honor by Dom Pedro, Emperor of Brazil. But Gray continued to downplay the success, saying, "as a scientific toy it's beautiful" but its "commercial value will be limited."

Bell offered to sell his telephone patent to Western Union for a pittance—$100,000. In legendary narrowsightedness, the giant telegraph company said no. (Two years later, Western Union would gladly have bought the patent rights for $25 million.) Instead, it paid a huge sum for Gray's multiple telegraph.

FIGHTING FOR THE PATENT

Doing a dramatic about-face the next year, Western Union tried to wrench from Bell what it had been offered the year before. In 1877 it backed Gray in a huge lawsuit seeking to strip Bell of his patent. While everyone agreed that Gray was an able and diligent scientist who had independently arrived at the same conclusions as Bell, no amount of Monday morning quarterbacking could convince the courts that Bell didn't deserve the patent. After a long court battle, Bell and Gray settled out of court.

The attitudes that marked the battle between Bell and Gray reflect the differences between entrepreneur and businessman. To

Gray, following the tempo set by his business partner, the telephone was an annoying curiosity—a hobby. To Bell, an amateur working virtually alone, the telephone was his life.

DID HE SPILL THE ACID OR DIDN'T HE?

Is there any accident more famous than Bell's spilling of the acid, which led to his shouting for Watson's help? Yet there is a good deal of doubt whether the whole affair ever happened. The widely accepted turn of events goes like this:

Bell is standing over his acid-filled transmitter, tinkering with the wires. Watson, standing in another room across the attic on Exeter Place in Boston, is waiting for Bell to send a musical signal through the wires linking the rooms.

As Watson waits in the distant room, Bell—a bit irritable after a long day of experimentation—accidentally bumps into the liquid transmitter, spilling the acid all over his pants. In a fit of utter disgust, he cries out for help in cleaning up the mess: *"Mr. Watson—Come here—I want to see you."*

An astonished Watson hears the words come out of his receiver. They are low in tone but clearly understandable. In a flash Watson is flying down the hall, beside himself with excitement. "Mr. Bell," he bursts in, "I distinctly heard every word you said!"

Though probably burned by the acid Bell ignores his ruined pants. Both realize they have made history. The first words ever carried over wires by electricity signal the age of the telephone.

The problem with this Hollywood-like sequence is twofold: One, the invention of the telephone was not an accident; the words spoken were deliberate. Two, the incident probably did not involve the spilling of any acid.

NO SPILL

The experiment conducted that day, March 10, was designed to test out a crude new telephone transmitter. Bell's telephone sketches submitted to the patent office three days earlier had used a solid voice coil for the transmitter. In this experiment, Bell was trying to improve upon the mouthpiece by substituting a liquid transmitter—acid—for the solid voice coil in the patent. Bell's handwritten notes describe what happened next.

> I then shouted into M [mouthpiece] the following sentence: "Mr. Watson—Come here—I want to see you."

Nowhere in Bell's notes written two days later or in Watson's own notes written that night is there any mention of acid being spilled. And again, years later, while giving sworn court testimony about the event, Watson failed to state that Bell's call to Watson was an inadvertent cry for help.

Only after Watson published his autobiography, fifty years later, did he describe Bell's call to him as resulting from spilled acid.

Why the conflict? If he did spill some acid, Bell, who was not mechanically inclined, might have been too embarrassed to tell anyone about it. Supporters of the spilled-acid account also claim that both Watson and Bell were poor record keepers; they failed to write down everything that happened, including the spill.

However, Bell's biographer, Robert V. Bruce, has another plausible explanation. He writes that Watson "probably confused that occasion with another, more satisfyingly dramatic."[2]

WHO WAS WATSON, ANYHOW?

When Bell decided to pursue the telephone, he approached America's greatest electrical genius, the aging Joseph Henry, who was close to death. Bell said he had an idea about sending voices by wire but didn't have the knowledge about winding coils. Henry looked at him and simply said, "Get it."

Bell's dilemma was typical of the times. The nineteenth century in America was an era of tinkerers: amateur inventors who in their spare time dreamed up amazing and useful contraptions but lacked the skill to build their projects from sketches. They sought the help of professional craftsmen and artisans employed by workshops that catered to them.

The Charles Williams shop at 109 Court Street in Boston was one of these places. Williams specialized in electrical devices. One of the employees of the Williams shop was Thomas A. Watson, a man skilled in the use of his hands who could take an idea and make it a reality.

Watson first met Bell in 1874, when Bell rushed into the shop for a quick fix of an instrument he was making. As Bell brought in more work to the shop, Williams assigned Watson to the Bell account. Watson and Bell became associates and lifelong friends. In 1875, following Henry's advice, Bell enticed Watson—then twenty-one years old—to leave his job at the shop and join him. Little did Watson know that in March 1876 he'd become the world's most famous listener.

WATSON LEAVES BELL

On the heels of their successful telephone collaboration, commercial telephone service quickly followed, beginning in 1877. That year Bell went off on his honeymoon and left Watson as chief telephone repairman. By 1881 Watson was fed up with the telephone. At the ripe old age of twenty-seven, Watson retired from the phone business to start a new life. Or as it turned out, many new lives.

With pockets bulging from telephone patent royalties, Watson took off for a long European holiday. He married and became a gentleman farmer, only to find that the country life was not for him. Soon he was back in a machine shop in a Boston suburb.

Business was good, so good that in the 1890s he added a new line of work: shipbuilder. He began building destroyers for the navy. The turn of the century found him operating the largest shipbuilding business in the country. Shipbuilding, as anyone in the business will tell you, is as fickle as the tides that carry the vessels. In 1903 the

Watson Dressed in Costume. Watson, the man who heard the first telephone message, is shown dressed in costume as the clerk in *The Merchant of Venice*, Stratford-upon-Avon, May 1911. Following his famous career as Bell's assistant, Watson changed careers numerous times and even formed his own Shakespearean company.

tide of Watson's fortunes turned. The board of directors of the company he founded replaced him as president.

The unsinkable Watson started a new venture. He became a geologist. Having taken a three-year course at the Massachusetts Institute of Technology in geology and paleontology while he was president of the shipyard, Watson refocused his attention downward. He teamed up with another geologist, and together they explored Alaska and California, searching for mines that would yield valuable ore deposits. But none turned up.*

---

* Watson was so good at his newly found avocation that fellow geologists named a fossil gastropod after him.

THE THEATER BECKONS

Out of work at the age of fifty-six, Watson redirected his career toward the stage. Recalling how he had loved studying the human voice under Bell's tutelage, Watson decided the life of Shakespearean acting was for him. In 1910 he joined Frank R. Benson's Company of Shakespearean Players in England. Given the typical role of a novice player—a face in the crowd—Watson enjoyed every minute of it. The next season found him with a speaking part at Stratford-upon-Avon, and by season's end, Watson was so invigorated by Shakespeare that he and a few players formed a road company of their own.

This renaissance man went on to adapt Dickens's novels to the theater, writing stage versions of *A Tale of Two Cities*, *Oliver Twist*, and *Nicholas Nickleby*.

Slowing down a bit by the age of fifty-eight (in 1912), Watson returned to his home in Braintree, Mass., to a life of amateur play production and public speaking. His favorite subject was the invention of the telephone. Recalling, in 1913, the highlights of his career, Watson said, "My greatest pride is that I am one of the great army of telephone men, every one of whom played his part in making the Bell Telephone service what it is today."

It was during one of the numerous interviews and talks he gave decades after the famous telephone event that he first mentioned that Bell had accidentally spilled acid on his trousers that fateful night.

Decades later Watson and Bell reenacted their famous discovery, but with a twist. During the inauguration of the first transcontinental telephone service in 1915, Bell sat in an office in New York and telephoned Watson in San Francisco. This time, in response to the famous "Mr. Watson—Come here—I want to see you," Watson replied, "I would be glad to come, Mr. Bell, but it would take more than a week." The call reportedly took twenty-three minutes to be put through at a cost of $20.70. Some things never change.*

* The first telephone exchange was opened in New Haven, Conn. in 1878. The first telephone operator, George W. Coy, greeted each caller with a hearty "Ahoy." It worked for boats, why not phones?

9

Who Really *Invented Television?*

Philo T. Farnsworth
Father of Television

— Inscription under the statue
erected by the State of Utah, U.S.
Capitol Building, Washington, D.C.

Vladimir K. Zworykin
Father of Modern Television

— The New Encyclopedia Britannica

John Logie Baird
Father of Television

— RONALD TILTMAN, television
historian

Success has many fathers.

— JOHN F. KENNEDY

IF YOU WANT to get into a good argument, start a discussion about who invented television. Try claiming that an American invented TV and you'll get a counterclaim from the British.

The problem with pinning down the invention of television to any one person is that television as we know it has been evolving for over a hundred years. (And I use the word evolving because TV is still changing shape.*)

* Considering the advent of high-definition TV, digital transmission and reception, and other innovations, we haven't even come close to the form of higher life television may evolve into.

89

Inventors seeking to create improved facsimile machines in the late 1800s found a method of sending a crude picture via wire using light-sensitive photocells (see Chapter 7, "Fax: The Priest and the Pendulum"). Was this the birth of television? Purists argue that since the fax inventors didn't pursue the idea of sending a continuous stream of pictures that make up a TV image, TV's birth should be attributed to those looking to transmit *moving images*, not stationary ones.

SPINNING WHEELS

Television relies on the ability to rapidly send many images. If they can be sent fast enough (twenty-four per second), the brain can smooth out the flashing images and create the sensation of a smoothly moving picture.

That concept was grasped in Germany by twenty-four-year-old inventor Dr. Paul Nipkow of Berlin, who in 1884 received the first patent for a television system. Nipkow called his system an "electric telescope," and by today's standards, it would hardly be recognized as television.

Nipkow's system was based upon two identical motorized spinning disks, one in the transmitter and another in the receiver. Each disk had a spiral pattern of twenty-four square holes punched into it. At the "camera" end, light from the scene being televised passed through the holes in the disk.

The rapidly spinning disk broke the image up into lines, each atop another. As the disk rotated, successive lines were traced out. The whole picture could be scanned right to left, the number of lines equaling the number of holes.

Once through the holes, the light impinged upon photocells that changed the light into an electrical signal. Carried by wires to the receiver, the signal was fed to a neon light. The intensity of the neon light changed in accordance with the signal. To reconstruct the image, the neon light was sent back through the holes of an identical spiral disk spinning in synch with the one in the transmitter. The light was projected onto a ground-glass screen where the original scene appeared. It was not a clear and clean image that the family could crowd around and enjoy. But it was a beginning.

A NEON BREAKTHROUGH

A new, brighter neon lamp, invented in 1917, brought new practicality to Nipkow's invention. It was just what American inventor Charles Francis Jenkins had been waiting for. Experimenting with the disks throughout the 1920s, Jenkins developed a small mechanical television system. Bought by hobbyists for the home, Jenkins's TV sets, complete with finished wooden cabinets, made exquisite television furniture. About two dozen TV stations transmitted mechanical television pictures in beautiful orange and black.

Nipkow's invention also helped launch the TV career of one of England's most intrepid inventors: John Logie Baird (see Chapter 10). Working independently of Jenkins an ocean apart, Baird constructed a TV system based on Nipkow's disk. His model was also very crude. Composed of only thirty lines and flashing at ten times per second, the pictures were barely watchable. They flickered badly and when viewed on a small, dim screen, they created nauseating headaches in viewers.

But Baird continued to improve his system to the point where it worked well enough to create a demand for television. By 1930 Baird was selling TV sets to the British public for about $130 apiece. Baird waited for regular daily radio broadcasts to finish each evening so that he could borrow the BBC radio equipment and transmit his video signals. Baird's success spurred research in England and led to pioneering TV broadcasts there between 1927 and 1935.

Though variations on the theme were developed by Baird and

other inventors around the world, the rotating disk scheme was not radically improved upon. And unless somebody did come up with something better, television would be dead commercially. Besides being so unwieldy, the spinning disks required bathing the scene in intense light. Humans just couldn't take the heat. Most of the light was wasted because only a fraction of it ever made its way through the holes.

A SANDWICH SOLUTION

Obviously, spinning disks were not the answer. Mechanics had to be done away with. Ironically, the solution would not come from the world of broadcasting. It lay in technologies pioneered in medical electronics. The architect of electronic television was a British electrical engineer, A. Campbell Swinton. He conceived his blueprint for television of the future in 1908 and laid it out in a speech to the Röentgen Society in 1911.

Swinton suggested focusing the televised image on a special electronic "sandwich." A nonconducting plate made of mica would be coated on one side with metal. The other side would be dotted with small "islands" of photoelectric material—metal that spits out electrons when hit by light. The sandwich would be placed inside a special cathode-ray tube invented by the British physicist William Crookes in the mid-nineteenth century (to produce X rays).

The image would be focused on the front of the plate—the island side. When struck by light, each photoelectric island would give off a negatively charged electron, leaving the island positively charged. The more light that struck the island, the greater the positive charge. In effect, the image would be stored on the plate as a series of charged (island) particles, varying in intensity in accordance with the intensity of the light falling on it.

Meanwhile on the back side of the sandwich, an electron beam swept back and forth across the metal plate, "reading" the charges of each island. The sequence of charges on the plate would be converted to a video signal representing the image to be transmitted back to the television receiver.

What about the receiver? Swinton had that figured out, too. He aimed to slightly modify the cathode-ray oscilloscope invented by Karl F. Braun of the University of Strasbourg in 1897. At the core of the oscilloscope was a cathode-ray tube. In the neck of the tube, an electron gun shot electrons at a screen coated with phosphor that glowed when hit. Magnetic coils surrounding the tube would sweep the electron beam back and forth across the screen, from top to bottom. The coils would be hooked up to the video signal coming from the camera so that movement of the electron beam in the camera matched the movement of the beam from the gun in the picture tube. The video signal would control the electron beam from the gun in the receiver and mimic the patterns of light and dark being seen by the camera.

Where the camera recorded brightly lit areas, the beam was fully on and the tube's screen would glow brightly. Darker areas meant the beam was partially turned off, producing gray. Black images shut the beam off entirely, producing corresponding black on the picture tube.

PUSHING THE LIMITS OF MECHANICAL TELEVISION

Swinton laid out before those assembled an electronic TV "shop manual." It was elegant and workable, and it literally formed the backbone of modern television. It was an idea way before its time, and even Swinton doubted anyone's ability to make such a system, saying, "I do not for a moment suppose that it could be got to work without a great deal of experimentation and probably much modification."[1]

The main drawback to mechanical television was its lack of picture definition. The only way to make the picture sharper was to increase the number of lines. Swinton had pointed out that for good resolution television pictures needed to be made of between 100,000 and 200,000 picture elements—discrete points of light. Mathematically, that boiled down to about three hundred lines of television resolution. But the mechanical TV disks could deliver at best about two hundred lines. Mechanical television was coming to a dead end. Swinton's electrical television began to look better and better to those who dared think about it.

THE YOUNGEST FATHER OF TELEVISION

Picking up the gauntlet was a young Mormon teenager who had toyed with the marvels of electricity while electrifying his mother's hand washing machine. Philo T. Farnsworth, an avid reader of electrical journals, had studied the Nipkow system and declared it dead. In 1922 the thin, frail-looking adolescent from Utah told his high school teacher that he could devise a better system, an all-electrical system. After laying out the system on the school blackboard, Farnsworth stood back as his teacher, aghast, told him to "go for it."

Traveling to California under the guidance of a financial backer, George Everson, Farnsworth set up secret workshops in Los Angeles and then San Francisco. On one occasion, his suspicious carryings-on—drawn window shades and mysterious glass tubes—drew the attention of the police, who raided his lab expecting to find a Prohibition-era still.

This clandestine work bore fruit in 1927 when he successfully transmitted a series of images including, fittingly, a dollar sign. The sixty-line picture was reportedly breathtaking. According to television historian Erik Barnouw, Everson declared the image "jumped out at us from the screen." Then followed film clips of boxing (Dempsey vs. Tunney) and Mary Pickford combing her hair in *The Taming of the Shrew*.

Farnsworth claimed to have solved both the transmitter and the receiver problem. This young upstart had designed a total system that did away with spinning disks. He immediately applied for a patent. With the world searching for a new television idea, so bold an action did not go unnoticed. It drew the attention of one of the

Early Electronic Television. Philo T. Farnsworth with his CRT receiver, circa 1929.

country's leading entrepreneurs, David Sarnoff, head of the Radio Corporation of America (RCA). Sarnoff welcomed Farnsworth and his invention by trying to buy him out.

THE GURU OF RCA

Sarnoff had already been talking up television. In 1923 Sarnoff told RCA's board of directors of his vision that someday each and every American home would be equipped with a television that "will make it possible for those at home to see as well as hear what is going on at the broadcast station."

Sarnoff saw Farnsworth as competition that must be neutralized at all costs. His lawyers at RCA first sought to discredit the potential patent. They challenged Farnsworth, peppering him with questions for hours. Farnsworth hung tough. Then they challenged Farnsworth's patent application, saying it infringed on work being conducted by RCA.

Already working on television at RCA was a brilliant Russian immigrant who had settled in the United States in 1919, just eight years after Swinton's challenge. Vladimir K. Zworykin, an electrical engineer in St. Petersburg, took any job he could find, including bookkeeping at the Russian Embassy. In his native Russia he had worked under the tutelage of Professor Boris Rosing, a physicist who believed as Swinton did that television could be done best using electrical, not mechanical, means.

ZWORYKIN AND HIS TV TUBES

Shortly after Christmas 1923, while working as an engineer for Westinghouse, Zworykin filed a patent for a crude television camera, claiming it was the first device to scan a picture entirely by electronic methods. But the patent office did not grant his patent (not until exactly fifteen years later, in December 1938). Eager to show off his work to his employers, Zworykin presented his crude electronic television system to Westinghouse brass in 1924. They were not impressed. The picture was not clear. The images were muddy, looking more like shadows than objects. A lot of work needed to be done.

Undaunted, Zworykin went back to the drawing board. Appearing once again before Westinghouse executives in 1929 with an improved version, Zworykin found another cool reception. Zworykin was having trouble perfecting his system.

SARNOFF TO THE RESCUE

What Zworykin needed was a visionary. Someone who could see the gold in the nugget of his idea and mine it. Someone willing to take a chance. What Zworykin needed was David Sarnoff.

Sarnoff, who had arrived in the New World as a Russian émigré himself, arranged for a meeting. Discussing the prospects, Zworykin speculated that it might cost about $100,000 to iron out the kinks in his new system. Sarnoff would wind up spending $50 million. He brought Zworykin on board at RCA in 1930 and put him to work perfecting the system, a device to be called an iconoscope—from the Greek *eikon*, "image," and *skopon*, "to watch."

In August 1930, at the age of twenty-four, Farnsworth received a patent for an electronic camera tube, called the dissector, which in some ways was better than Zworykin's yet-to-be patented idea. RCA had to find out more about it. Out to the coast came Vladimir Zworykin to have a look.

Arriving in California, RCA's television expert examined the youngster's invention and declared that there was nothing that RCA would be interested in. Skeptics claimed that Zworykin saw a lot more of interest than he admitted publicly. (Years later they would accuse him of stealing ideas.) His curiosity piqued, Sarnoff soon followed and echoed the sentiment. Farnsworth had nothing they couldn't live without.

But RCA's lawyers thought otherwise. They knew trouble when they saw it. Farnsworth's patent could cost them lots of money in the years ahead. If television did become as popular as they hoped, RCA might owe Farnsworth loads of money in royalties.

Farnsworth was more than eager to negotiate a royalty deal. He might have been young, but he wasn't stupid—remember, the dollar sign was the first TV image he had transmitted. He insisted RCA pay him royalties.

But RCA wouldn't hear of it. As historian Erik Barnouw put it: "This was a company used to collecting royalties, not paying them." For more than a decade RCA hounded Farnsworth, trying to buy him out or grind him down. But to no avail. Finally, in 1939, RCA acquiesced and agreed to pay royalties. For the first time in its existence, RCA was forced to buy a license. "The RCA attorney is said to have had tears in his eyes as he signed the contract," wrote Barnouw.[2]

EXPERIMENTAL AMERICAN TV

Left with no marketable alternatives, American companies continued developing mechanical television. In April 1927 AT&T showed off its latest design by transmitting pictures of Secretary of Commerce Herbert Hoover two hundred miles by wire from New York to Washington. Just days later AT&T topped that feat by transmitting TV signals by radio for the first time between New York City and Whippany, N.J., about twenty-seven miles apart. General Electric jumped into the act by telecasting a play in Schenectady, N.Y., in 1928.

Even RCA was keeping its options open and its pockets deep. Realizing the intensity of the race for TV, Sarnoff continued to explore mechanical television while Zworykin struggled with the electronic side. RCA tried combining the two systems, testing out a hybrid, but it failed.

One of RCA's first scheduled television shows was broadcast on Tuesday evening, July 21, 1931. Kate Smith made her TV debut, singing "When the Moon Comes Over the Mountain."

ZWORYKIN COMES THROUGH

In 1933 Zworykin came back to the boss at RCA and boldly claimed that his electronic television was ready for the public. His system consisted of an improved iconoscope as a camera and, for the picture tube, the kinescope, based on the Braun oscilloscope. Zworykin improved the camera by finding a way of shrinking the photosensitive islands so that a million of them could be put on the camera plate.

Was this the breakthrough RCA had waited for? Zworykin's system could work under normal studio and outdoor lighting conditions, with better picture detail than before. Why not put it through a full test? Setting up a receiver four miles away in Collingswood, N.J., RCA successfully transmitted an image of Mickey Mouse (among others) to a nine-inch picture tube. A mirror reflected the image to the front of the set.

It worked. Electronic television of high enough quality to be commercialized. Just a bit of tweaking here and poking there, improve the resolution a bit and get the bugs out, and the tube should be ready for the public.

EUROPEAN TV FIRST

Europe certainly thought so. Some of the best research labs in Europe were ready to challenge RCA's electronic TV. Germany's Manfred von Ardenne designed an improved picture tube, and by 1935 regular broadcast service was started in Germany with a medium resolution of 180 lines. The British Electric and Musical Industries (EMI) wisely reasoned that no television camera or receiver would be of any use without a complete and standardized support system. EMI put a whole TV system together, based on an improved iconoscope called the Emitron. That advanced camera tube, patented in 1932, coupled with an improved cathode-ray tube for the picture, formed the backbone of a British system that would last for decades.

By 1935 EMI put a complete TV package together: cameras, tubes, amplifiers, switchers. Seeking to uproot the TV beachhead held by Baird, EMI ran comparison tests of its new system with Baird's mechanical box. In 1936 BBC viewers were treated to one week of EMI television alternating with one week of Baird TV. Baird's improved 240-line, twenty-five-frame system challenged the EMI electrical system for a while but shortly gave in. Baird proved no match for EMI's picture quality. Under the leadership of Isaac Shoenberg, EMI set the technical standard for resolution. Shoenberg proposed the British use a picture made of 405 lines, flashing at fifty interlaced frames per second, yielding a picture of twenty-five frames per second without flicker.

The British government liked the idea; it gave EMI the thumbs-up. This was the world's highest definition television. Inaugurated in London in 1936, the service was so good that it lasted until 1964.

In 1938 the first regular television program schedules were published amid the radio schedules in *Radio Times*.

VIABLE TELEVISION COMES TO AMERICA

A year later, 1939, saw RCA ready to unleash TV to the American masses. Its RCA Exhibition Hall at the New York World's Fair featured television as its centerpiece. But the high-priced publicity could do little to overcome American TV's two great shortcomings: price and politics. The World's Fair succeeded in putting a few hundred TV sets into the homes of wealthy New Yorkers in 1939. But the price tag of $625 each was out of the reach of most Americans. And just as well, because RCA's license to broadcast television was for "experimental" purposes only; no commercial service was allowed. In addition, the RCA system would not be the one finally adopted as the standard broadcasting system in the United States. In 1941 the U.S. Federal Communications Commission established the standard for American television: 525 lines per frame, thirty frames per second.

In the meantime improved versions of the television tubes came along, beginning with the improved dissector tube of Philo Farnsworth in 1934 and the orthicon by RCA in 1939.

The war years severely curtailed the commercialization of television. Just a handful of sets were sold. Zworykin and RCA took that time to improve the camera tube and came up with one a hundred times more sensitive than the iconoscope that allowed TV cameras to peer into the shadows.

The end of the war saw the baby boom, Uncle Miltie, and the explosion of television. In the fall of 1946 a ten-inch table model could be yours for $375. The era of the couch potato had begun. Radar, which had been such a hero of World War II, was a close cousin to television. Consumer TV technology, which had been put on hold during the war, found a boost from all the radar experts who found themselves without jobs once the war was over. Consumer TV repair shops found a ready supply of TV technicians who had previously been radar experts.

SO WHO INVENTED TELEVISION?

If you're British, your hero is John Baird who gave the first true (?) demonstration of (mechanical) television in 1926 by transmitting half-tone pictures. Or you could point to the 1936 high-definition* London broadcasts. If you're American, you'd say Zworykin and point to his 1923 patent application.

If you're a keen student of history (or from Utah), you might say Farnsworth, for receiving the first patent for electronic television.

The issue of who invented electronic television was put before the courts in a famous court battle of 1932, *Farnsworth v. Zworykin*. In this patent interference suit, one court after another acknowledged that Farnsworth should be considered to have beaten RCA to the electronic TV punch.

FARNSWORTH: TELEVISION'S UN-PERSON

Despite Philo T. Farnsworth's early success in pioneering television, it's hard to find him mentioned in a popular modern history of TV. He seems to have disappeared—or perhaps to have been made an "un-person." Browse through the library stacks and see how many books on the history of television devote much ink to Farnsworth.

For someone who had received a patent for inventing television, his name hardly surfaces in the literature. For example, in an article written in the respected (now defunct) science magazine *Science 84*, he isn't even mentioned in a key piece about the history of television.

Grolier's *Electronic Encyclopedia*, 1990 (on computer disk), doesn't mention him at all. According to *Channels* magazine writer Frank Lovelace, the *Encyclopedia Americana* "virtually dismisses

---

* High definition in this sense is not to be confused with the current day-meaning of high definition TV. For 1935 high definition was 525 lines. For the 1990s high definition is upwards of 1,000 lines.

Farnsworth" in an article about the development of television written by Edward G. Ramberg. Ramberg was employed by RCA at the
time and co-authored with Zworykin a book about television.

The *New Encyclopedia Britannica* gives Farnsworth a brief biography. European publications do better. Many more acknowledge
the existence of Farnsworth. And the older the book or article, the
more his name appears.

Cynics claim the *1984* job being done on Farnsworth is a direct
outcome of RCA's bitterness after losing the 1932 lawsuit. Since
RCA engineers have written lots of the articles about the history of
television and since RCA promotional materials barely mention the
contributions of Farnsworth, Farnsworth supporters claim RCA is
out to erase his name from history. In the TV wars the history of
television has been written, more or less, by the victors.

Before his death in 1971 Farnsworth enjoyed an illustrious professional career. He invented a fistful of video devices, including
amplifier and cathode-ray tubes and scanners. He founded the
Farnsworth Radio and Television Corporation in 1938 and helped
develop radar. Farnsworth even contributed to the developing
technology of nuclear fusion.

WHY YOU CAN'T WATCH YOUR AMERICAN TV IN EUROPE

Rival American and British TV systems still haven't kissed and
made up from their rivalry of the '30s. European and American
color TV sets are incompatible. You can't plug in a standard American color television in Europe and expect it to work. And vice
versa. The TV sets are as foreign to one another as the power
cords are—and for exactly the same reason.

The television camera in the studio and your set at home have to
be synchronized. Just as moving picture film is a series of rapidly
moving frames, so are TV pictures, transmitted swiftly one at a
time. For your set at home to receive the pictures in synch with the
camera taking them, engineers had to come up with some universal

timing device that would be the same in your home as it is in the TV studio.

Instead of shipping everybody a clock in their TV sets, they hit upon using something everyone already has: electric house current. If you live in the United States, your house current alternates—reverses direction—at a rate of sixty cycles per second. Television engineers decided to synchronize the television picture frames according to the frequency of the alternating current, the same for both camera and TV set. The TV camera generates one frame for each two cycles of AC, yielding an American standard of thirty frames per second.

But in Europe and Great Britain, the AC runs at fifty cycles per second, yielding a picture generated at twenty-five frames per second. Hence the incompatibility of the two standards. To make matters worse, the number of lines per picture are different in the two systems. British and Europeans enjoy a better picture because their system is made of 625 lines. American television is only 525 lines. Being quite incompatible, American and European TV systems cannot make pictures out of each other's systems. And they can't share videotapes made by each other.*

Where is all of this heading? It's too late to change the present system. Perhaps the solution will lie with high-definition television. HDTV may force everyone to eventually purchase new equipment anyhow. Will the HDTV standards of the future make TVs compatible worldwide? Tune in tomorrow.

---

* The American system is called NTSC, devised by the National Television System Committee of the Federal Communications Commission in 1941. The Japanese system is also NTSC (as are those of Canada, Mexico, and some South American and Asian countries). Western Europe, Australia, and some South American and some South African countries use the PAL (phase-alteration line) standard. France, various East European countries, and the former Soviet Union (now the Commonwealth of Independent States) adopted the SECAM standard (*sequentiel couleur avec mémoire*, or sequential color with memory). But before you lug your portable color TV abroad, be advised that at least thirteen variant subsystems of the three standards exist around the world. As they say, check your local listings.

10

······································

John Logie Baird: The Scotsman Who Almost Won the TV Race

Whatever may be the future of television, to Baird belongs the success of having been a leader in its early development.
—*New York Times*, February 28, 1928

How DO YOU know when you've become a successful television broadcaster? When you have a series of television programs coupled with a way of transmitting them, an audience that can see them, and a way of letting people know when to watch.

John Logie Baird had all of them. Baird was England's television pioneer. By the early 1930s Baird was transmitting an eclectic array of TV fare, from a stage play to a horse race. And for many years, he successfully broadcast TV programs to homes in Britain while his competitors in the United States waited for television to be developed.

At the age of thirty-four, Baird had been a miserable failure as a businessman and a victim of chronic poor health. Having failed in both the soap and the jelly business, Baird turned his thoughts toward the rage of the age: finding a way of sending pictures by wire and radio. Baird saw the new neon light and thought it was good. Working out of an attic room in London's Soho on a budget of £250 raised from his father and a business friend, Baird made his own crude television system based on the Nipkow disks. This was no high-tech adventure. His parts came from objects found in the attic: An old tea chest formed the base to carry the motor. An empty biscuit box housed the lamp. "Scanning discs were cut out of cardboard," wrote eyewitness and friend Ronald Tiltman, "and the mountings consisted of darning needles and old scrap timber." Complete with an old bicycle light lens, a couple of hatboxes, and endless lengths of wire, the entire rickety affair was shakily held together with bits of sealing wax, glue, and lengths of string. For months, Baird tried to coax a picture out of the contraption. But it refused to cooperate.

Finally, one afternoon early in 1924, after relentless tinkering, Baird got his contraption to hold together long enough to successfully transmit the tiny pink, flickering image of a Maltese cross over the distance of a couple of yards. Baird's television produced neon-pink pictures with stripes wide enough to almost hide the image. But they were pictures, nonetheless. And they were the first anyone on the continent had produced.

John Logie Baird's Televisor. Baird was the first to establish a working television system for public use. As shown here, Baird's first experimental images produced in the early 1920s were crude affairs—doll-like cutouts.

THE WORLD'S FIRST TV STAR: A DUMMY

In 1925 the retail store Selfridge's caught wind of Baird's invention and paid him twenty pounds per week to demonstrate it in the store. Publicity stunt or not, the Scotsman jumped at the idea. Three times a day, long lines of people queued up to see the dark images of the first TV pictures televised a few feet away. Improving his apparatus, Baird was able to televise the world's first TV star: a ventriloquist's dummy named Bill. Elated, Baird turned his attention to the reluctant face of the office boy, William Taynton. Let history show that the first human being ever to be televised had to be bribed. Buoyed by his success, Baird exchanged places with Taynton who, as Tiltman wrote, "became the second person in the world to see a televised human face."

Buoyed by his three-week run at Selfridge's, Baird decided to throw a party. In 1926 he invited forty members of the Royal Institution for a demonstration. Arriving in full evening dress, the guests lined the narrow staircase to his dilapidated laboratory, hoping to catch a glimpse of the apparatus. Everyone went home satisfied, including the press. The next day's *London Times* carried a glowing account of the event and Baird, who couldn't rub two cents together, became an overnight sensation.

Financial backers opened their wallets, allowing the poor and ailing Scotsman to eat and live well for the first time. The Post Office approved, without trouble, a license for transmitting pictures—the world's first. Baird was becoming a legend in his own time—on the continent. But from his American competitors he received nothing but snubs. In 1927 AT&T would stage a show called "Television at Last" and ignore Baird's work. Baird would make sure that didn't happen again.

BESTING THE YANKS

When in 1927 AT&T broadcast pictures 200 miles from New York to Washington, Baird cranked up his set and transmitted pictures

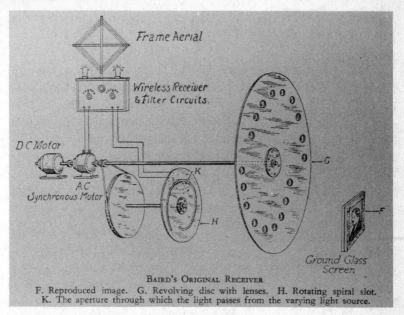

BAIRD'S ORIGINAL RECEIVER
F. Reproduced image. G. Revolving disc with lenses. H. Rotating spiral slot.
K. The aperture through which the light passes from the varying light source.

Schematic drawing shows how simple televisor was constructed.

from London to Glasgow—430 miles. When AT&T successfully transmitted mechanical pictures by radio between cities in New York and New Jersey twenty-seven miles apart, Baird topped that feat in 1928 by beaming the world's first mechanical TV signals across the Atlantic to the United States.

The *New York Times* of February 11, 1928, proclaimed in ringing tones that such an achievement put him in a league with Marconi, the first man to send wireless radio signals across the Atlantic. "All the more remarkable is Baird's achievement," said the *Times*, "because he matches his inventive wits against the pooled ability and the vast resources of the great corporate physicists and engineers, thus far with dramatic success. Whatever may be the future of television, to Baird belongs the success of having been a leader in its early development."

Baird also investigated the technology of seeing in the dark—night vision—with an infrared device called Noctovision. He designed a mechanical color television based upon a Nipkow disk that has three sets of holes, each covered with gels of the primary colors.

Baird's mechanical systems worked to his satisfaction. On September 30, 1929, the BBC made its first experimental broadcast. At eleven in the morning an announcer came on the screen and read a letter. It was followed by a speech by the inventor of the vacuum tube, Sir John Ambrose Fleming. The image was viewed on the TV set called a televisor, and its quality was described as "similar to that of looking into an automatic picture-making machine as installed in music halls." The image appeared to be "as a soft-tone photograph, illuminated by a reddish-orange light."

Public Television. Reproduction of advertisement from Selfridge's Department Store. Baird's TV was put on public display for three weeks in 1925. A ventriloquist's dummy became the first TV star, followed by Baird's office boy, who sat in front of the camera only after being promised a small reward.

SELFRIDGE'S

Present the First Public Demonstration of Television in the Electrical Section (First Floor).

Television is to light what telephony is to sound—it means the *INSTANTANEOUS* transmission of a picture, so that the observer at the "receiving" end can see, to all intents and purposes, what is a cinematographic view of what is happening at the "sending" end.

For many years experiments have been conducted with this end in view; the apparatus that is here being demonstrated, is the first to be successful, and is as different to the apparatus that transmits pictures (that are from time to time printed in the newspapers) as the telephone is to the telegraph.

The apparatus here demonstrated is, of course, absolutely "in the rough"—the question of finance is always an important one for the inventor. But it does, undoubtedly, transmit an instantaneous picture. The picture is flickering and defective, and at present only simple pictures can be sent successfully; but Edison's first phonograph announced that "Mary had a little lamb" in a way that only hearers who were "in the secret" could understand—and yet, from that first result has developed the gramophone of to-day. Unquestionably the present experimental apparatus can be similarly perfected and refined.

It has never before been shown to the Public. Mr. J. L. Baird, the sole inventor and owner of the patent rights, will be present daily while the apparatus is working—in the Electrical Section at 11.30 a.m., 2.30 p.m., and 3.15 p.m. He will be glad to explain to those interested in details.

We should perhaps explain that we are in no way financially interested in this remarkable invention; the demonstrations are taking place here only because we know that our friends will be interested in something that should rank with the greatest inventions of the century.

SELFRIDGE & CO., LTD.

Baird improved it by substituting a drum full of mirrors in place of the spinning disk. By the 1930s he was transmitting a wide variety of programs. On June 3, 1931, he sent his truck-mounted camera to Epsom Downs to televise the Derby. "The results astonished us all," shouted the *Daily Herald*. "We could see the horses passing in file; we heard them named by the announcers as they passed. We could almost recognize their jockeys."

Turning his televisor toward the arts, Baird broadcast legitimate theater, *The Man with a Flower in His Mouth*, among his early successes. Baird built and sold one thousand televisors at eighteen pounds each. Baird even had a regular TV slot: Each night, starting at 11 P.M., viewers could tune in his programming.

Baird's reputation was reaching America. The *New York Times* of September 13, 1931, included his television images sent by wire among "The Outstanding Inventions of the Past Eighty Years."

But he reached a point where if he were to go further, he'd need the backing of the only real player in the British broadcasting game: the government-owned British Broadcasting Corporation. The BBC owned all of radio and would certainly control the fledgling television industry. Baird would have to get the BBC to adopt his mechanical standard if TV were to really take off.

BBC engineers were skeptical. Certainly his system worked as a novelty; one could see images. But the dim, flickering images were not of good enough quality to get BBC backing. Yet the politics of the situation did not allow the British government to come right out and turn down a national hero. Prime Minister Ramsay MacDonald even had a televisor of his own at Downing Street.

But the BBC could see that electronic television was posing a real threat. The heavy hitters like Westinghouse, RCA, and AT&T were now developing the electronic camera. Baird had improved his system from the early 30 lines up to 240 lines, providing a sharper, clearer picture. But EMI had developed an improved electronic camera called the Emitron. Trying to hold off electronic television in the mid 1930s was akin to fending off compact disks in favor of vinyl LPs. It was an approaching tidal wave that could not be stopped.

THE GREAT TELEVISION CONTEST

So the British government instructed the BBC to let the public decide which system it wanted. For one week, beginning in November 1936, Baird's TV system was shown, alternating with a week of Marconi EMI television. Baird proudly stood back and hoped that despite the marvels of the electronic age, a grateful public would reward a British hero his hard-fought-for prize.

It was not to happen. After three months, the experiment was stopped. The BBC chose the electronic system, took Baird off the air, and told him to pack up and go home.

Baird was heartbroken. He regretted having worked alone for so long and realized too late the wrong path he had taken. Had he chosen to join forces with another, major corporate player, he might still have been in the game. Or had the BBC not owned a monopoly on television, he might have started his own independent company.

Instead, he tried to sell his TV system to movie houses. But World War II nixed that idea when the BBC TV service closed up shop for the duration.

In 1946 John Logie Baird died at the age of fifty-eight. Forgotten by most of the world, his brilliant contributions to the development of television are almost unknown in the United States. He is certainly one of the tragic figures in twentieth-century technology. Baird's inability to change with the technology of the times left his genius untapped. But let's not overlook the dramatic success achieved by Baird in paving the way for postwar television, in awakening the public to the potential of television, and in giving them a fleeting taste of what might be.

First Death by Lightning Experiment. Attempting to repeat Franklin's sentry-box experiment, Russian physicist G. W. Richman was electrocuted when his device was struck by lightning. A foot-long spark jumped from his ungrounded wire to his head, killing him instantly.

The Forty-Hour Breakthrough. This artist's rendition shows the infamous light bulb that reportedly burned for over forty hours and proved to Edison that he'd found the secret to success. Edison is pictured using a battery to drive the last of the gases from the filament. Francis Jehl is pouring mercury in the reservoir of the vacuum pump. Jehl's handwritten note on the back of a copy of this print verifies the authenticity of the event. Francis Upton stands behind Edison with Charles Batchelor looking over his shoulder.

The First Electric Chair. This crude chair was used for the electrocution of William Kemmler on August 6, 1890, in Auburn (NY) Prison. Kemmler did not use the foot rest, which could slide back. Nor did he die after the first jolt of electricity.

Workers laying direct current cable under streets of New York.

Cartoon of George Eastman lugging his darkroom, 1877.

French advertisement for Brownie introduced in 1900 and selling for one dollar. It was originally designed for children, as indicated by this 1903 ad.

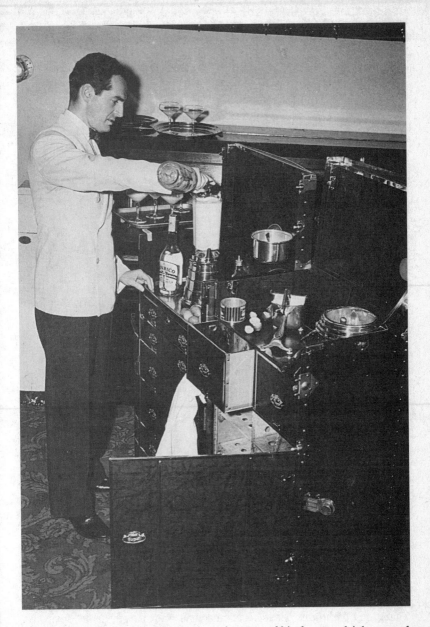

Bar in a Trunk. Fred Waring concocting one of his famous drinks out of his portable bar he carried in a special trunk while on tour with his band. Waring did not drink alcohol. He usually mixed fruit drinks for himself.

Swinging Pendulums. Caselli's fax machine was a large, heavy device. A swinging pendulum served as a time base for scanning a document. A clock, shown at right, served to synchronize the swinging of an identical pendulum equipped with a pen at the receiving end.

Early Televised Concert. Ruth Rowe, pianist, performs in studios of W2XCD, Jenkins Television Corporation, in New Jersey. The camera is a spinning disk plying spot scanner, circa 1931.

First Human TV Picture.
Human face as it appeared
on the receiving screen of
the original Baird televisor in
1926. Note series of
concentric images produced
by spiraling holes on disk.

Before Laser Beams there was. . . .the Photophone. Who needs lasers? Alexander Graham Bell invented the photophone 80 years before the laser. Sunlight bounced off a reflector and was focused into a narrow beam. The beam bounced off a shutter like mechanism that vibrated in response to speech. Speaking into the mechanism, Bell changed the intensity of the light beam in step with his voice patterns.

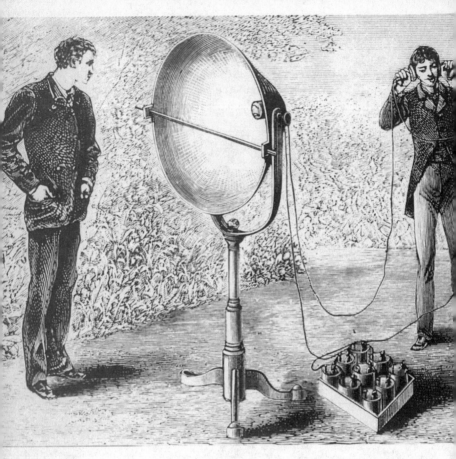

At the receiving end of the photophone, a detector made of selenium (at the back of the receiving "dish") caused electricity in the receiver to vary in intensity with the light. Hook a telephone up to the receiver and one could hear Bell on the other end.

Expensive Stockings. Betty Grable auctions her nylons for $40,000 at a war bonds rally. The nylon was recycled to make parachutes.

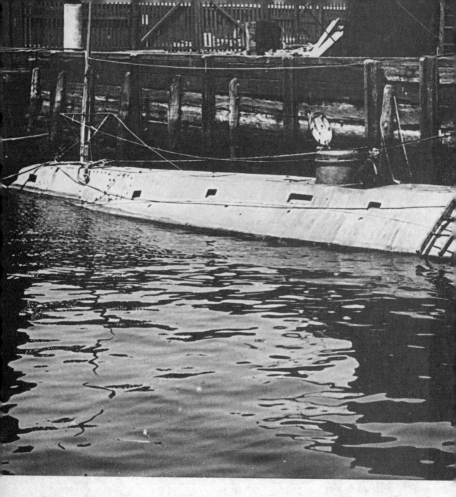

Defense Budget. The first submarine bought by the U.S. Navy (1890), the *Holland VI* was later christened the *USS Holland.*

A Dream Come True. From David Parkinson's dream came this rough sketch depicting an electrical control system for directing antiaircraft guns. The gun system was hurriedly built and proved decisive in shooting down aircraft over Britain during World War II.

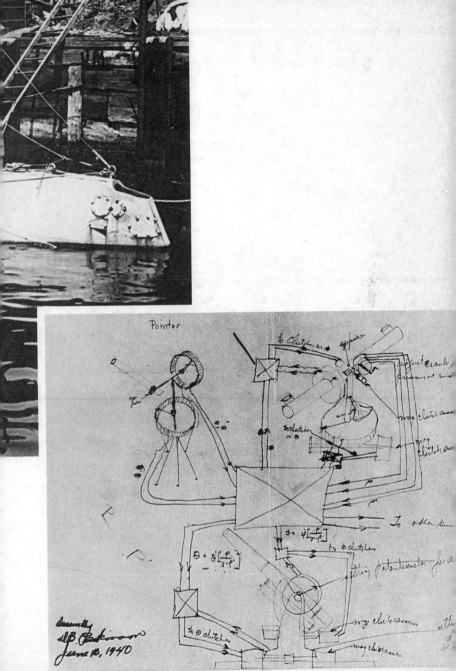

Pointer

Drawn by
J.B. Parkinson
June 13, 1940

Remington Notes

Volume 4   **Number 11**

Remington opened to women the doors of business life

The Big City Beckons. The cover of *Remington Notes,* circa 1907, graphically illustrated how the typewriter drew women to the office pool.

Xerox: The Machine No One Wanted

The most pleasure we get from life is sweating.
—CHESTER CARLSON, inventor of
Xerography, from his diary, 1928

CHESTER F. CARLSON was miserable. As a patent analyzer for an electrical component maker, he would spend long, tedious hours poring over patent documents and drawings. The patent office required multiple copies of drawings and documents, and the only way to duplicate them was to meticulously copy them by hand. And they had to be perfect copies. Redrawing took hours. On top of that, Carlson was nearsighted and his cramped writing posture made his arthritis unbearable. He was in agony. There had to be a better way. If there was one, Carlson swore he'd find it.

Carlson's tedious job analyzing patents was not a career choice. In college, in 1928, at the California Institute of Technology, he had found it difficult to choose a career. He was attracted to professions that would allow him to work in seclusion, away from social pressures and distractions. After toying with the idea of becoming a poet, a writer, and an artist, Carlson settled upon the life of an inventor. "I plan to . . . go fully into a project of organizing a business for inventing, research on inventions and the buying and development commercially of patent interests on or before the age of thirty."

Despite his burning ambition, Carlson fell victim to the times: the Great Depression. Graduating in debt, he couldn't find a job. Scores of companies rejected his applications. One that did not, Bell Telephone Laboratories, laid him off shortly after he was

111

hired. He wound up first analyzing patents for a law firm and then for P. R. Mallory & Company, an electrical parts maker. It was at Mallory that he recognized a need for a machine "that could be right in an office where you could bring a document to it, push it in a slot, and get a copy out." The idea for the photocopy machine was a child of the Depression.

AN IDEA BORN IN THE KITCHEN

Carlson set up shop in the kitchen of his apartment in Queens, N.Y. His main research source was the New York Public Library, where he pored over technical journals from Eastman Kodak, seeking the secrets of photographic duplication. But photography was too impractical to be applied to the office. It was very wet and messy. Photos had to be developed, fixed, and dried. The process involved too many chemicals, and a print could take hours to produce, if not days. This was not the path to follow.

Instead, Carlson had his own idea for a dry reproduction technique based on the principle of photo-conductivity. Certain materials, when exposed to light, change their electrical properties. In fact a Hungarian physicist, Paul Selenyi, had shown how charged particles would attach themselves to an oppositely charged surface. Here was the secret for making dry reproductions.

If Carlson could get dry particles to stick to a charge plate in a pattern corresponding to an image shining on the plate, he could make dry reproduction work. He called this idea electrophotography. Simply charge up a photoconductive metal plate with electrostatic charges. Project an image onto it. Where light from the image hits the plate, the charge disappears in total or is lessened, depending upon the intensity of the light. Dust the plate with oppositely charged powder, so it sticks to the oppositely charged areas. Now you have recorded the image on the plate in powder. Bake the image onto a sheet of paper and, voilà, you have made a dry reproduction.

It was a flash of genius. And like many great ideas it was simple and straightforward, looked great on paper, but proved difficult to execute. Carlson spent years concocting foul-smelling experiments in his apartment. His neighbors complained of the odor of rotten eggs. The landlady's daughter, as the story goes, came up to see what the sulfurous source of the odors was and wound up marrying Carlson in 1934.

Through the rest of the decade Carlson worked not only on his money-paying job but on his dream in the kitchen. There was no rest. When he thought that a law degree would help him get ahead in the office and protect his rights to his research, he entered New York Law School in 1936, worked every spare hour studying nights and weekends, graduated in 1939, and was admitted to the bar in 1940.

FROM KITCHEN TO BEAUTY SHOP

But success was slow in coming. Although Carlson received a patent in 1937 for electrophotography, he could not turn his dream into reality. He had yet to make his first dry print. It was time for a change. "Move out of the kitchen," said his wife, so he relocated to more luxurious digs at the back of a beauty shop owned by his mother-in-law in Astoria, Queens. Jumping in with both feet, Carlson went on a spending spree and hired an assistant, Otto Kornei. Physicist Kornei, a German refugee, was appreciative of the ten dollars a month research budget.

Then finally, on October 22, 1938, came the break they had been waiting for. To create a static electricity charge on a sulfur-coated zinc plate, Kornei rubbed it with a cotton cloth. Then he and Carlson held a glass microscope slide with words written on it in ink against the plate. The whole contraption was held under the light of a gooseneck lamp. After exposure to light for a few moments, the slide was removed and the plate dusted with powder (lycopodium; moss spores). Carlson pressed wax paper against the powder,

4

(Print I.)

Today Kornei & I made the attached permanent prints by the following methods: In both cases the print was made by using a 2 x 3 inch zinc plate having a thin layer of sulfur melted onto the surface. This was prepared by using a polished zinc plate (such as is used for etchings), etching the surface slightly in dilute nitric acid to form a matte surface and then melting CP crystalline sulfur onto it in a uniform layer over a hot plate. Then the sulfur surface was sandpapered with very fine glass paper in water to provide a smooth surface free from crystal structure. The plate used was marked on the back "Oct. 21, 1938 Ground S-layer on H NO₃ etched zinc". Thickness of sulfur .2 - .3 m.m.

Print I.— For making print I a thin sheet of celluloid was laid on top of the dry sulfur surface and the figures 1 2 3 were written on the celluloid with an ordinary glass rod. This produced a charge pattern on the sulfur surface.

(Print II)

The sulfur surface was then dusted with lycopodium powder and the 1 2 3 appeared in dust. The sheet of wax paper was pressed onto the surface with the hands and the dust transferred to it. The wax paper was then momentarily heated to melt the wax and fix the powder in the surface. •

10.-22.-38
ASTORIA

Print II.— Print II was made as follows: "The legend "10.-22.-38" was printed in India ink on a
ASTORIA
glass microscope slide. The sulfur layer was then charged by rubbing the surface with a linen handkerchief to produce a uniform background charge. This

---

The Famous Date. Chester Carlson's notebook showing slide with historic first successful photocopy: the date and location of his makeshift lab in Queens, N.Y.

heated the paper to melting point, and peeled it off. Blowing off the residual powder, Carlson could not believe what he saw.

Clearly duplicated on the paper were the words he had written on the slide: 10–22–38 ASTORIA.

DISBELIEF

Jubilant, Carlson believed that his three years of work had not gone unrewarded. Kornei was not so happy. To him, the words were disappointingly blurred. He could not see the word success hidden in the Queens dateline. Lacking the staying power and never-say-die attitude of great inventors, Kornei gave up and left to work for IBM. It was an omen of what was yet to come. Despite his initial success, Carlson could not get a company interested in his idea. IBM, GE, and RCA all turned away, refusing to sink any money into the idea.

Undaunted, Carlson built a wooden prototype of the machine and fiddled with it for five years. But to no avail. No one was interested. Carlson tried to get the idea out of his mind, he tried giving up on it, but it wouldn't go away.

What did go away was his money. Nearly broke and exhausted, he got his second break in 1944. Battelle Memorial Institute, a private nonprofit organization that invested in research, sent a man to P. R. Mallory to discuss other patents. Carlson pulled him aside and bent his ear about his own research. Battelle bit. It signed on with Carlson and agreed to invest $3,000 in research and development and act as Carlson's agent. In return, Carlson would receive 40 percent of the profits. (Eventually Battelle's share of the royalties would add up to about $350 million.)

But the good times didn't last for long. Carlson's wife, fed up with playing the role of a research widow, divorced him in 1945. Battelle started to run out of research money and beat the bushes for corporate funding. But again, no one was interested. The dry copier was about to run out of luck when a guardian angel appeared in the form of John Dessauer.

THE BIRTH OF XEROGRAPHY

Dessauer was director of research for the Haloid company. The small maker of photographic supplies in Rochester, New York, was in a fix. With its 1946 sales figures slumping and profits in a rut, it was in dire need of developing new businesses. Its old business, the manufacture and sale of photographic paper, was just not doing well. Dessauer had read an article about a new dry copying technology patented in 1937 by Chester F. Carlson. It could be the answer to his problem. But it would be a big gamble. The research would cost the company over $25,000 per year against a corporate income of just over $100,000. But Haloid took the chance. In 1946 it bought a license to develop a dry copying machine based on Carlson's design. The risk paid off. Tiny Haloid is now the Xerox Corporation.

Taking over, Haloid decided their product needed a better name. Electrophotography was just too cumbersome. A classic language professor at Ohio State University suggested the name xerography, from the Greek *xeros* for "dry" and *graphos* for "writing."

An uplifting name did not lift the spirits of Haloid's executives. Hardly anybody was optimistic that a useful product could be made. The dry copy process was a lemon, they insisted. The potential market for it was just a few thousand offices. Outside advisers said it would be insane to go ahead.

Nevertheless, the Haloid-Battelle team pressed ahead. Battelle refined and improved the method of fusing the image to paper. Haloid designed a stylish box for the machine. Ten years to the day after Carlson first held his glass slide to the light, the dry copy process was shown to the world. At the 1948 Optical Society of America meeting in Detroit, the public greeted the copier with a resounding *thud*.

Echoing the thoughts of Elisha Gray about Alexander Graham Bell's telephone, the critics said it was interesting but "had no future."

Determined, Haloid put its first machine to use xerography on the market in 1949: the XeroX Model A, popularly called the ox box. It was a bust. Crude and complicated, the ox box needed fourteen different manual operations to work. Consultants pointed out that the dry copier could hardly compete with the abundant and cheap supply of carbon paper. Besides, how many companies needed to make enough copies to justify the cost of a machine expected to be priced at around four hundred dollars. Perhaps you might sell a thousand or so to replace high-end offset printers, said early market estimates. But don't expect to have a machine in every office.

They were right, so far. The 1950s produced a partial success. In 1955 Haloid marketed an automated copier, the Copyflo, which produced prints from microfilm. It was enough of a success for the company to change its name to Haloid Xerox in 1958.

Although he made frequent trips to Rochester to keep tabs on the progress, Carlson was never hired by Haloid but remained an unofficial adviser. He lived in poverty off royalties from xerography and siphoned as much money as he could back to Haloid and Battelle for research. The payoff would come in 1960, more than twenty years after his invention.

MAKING AMERICA SAFE FOR TONER

Camelot came not only to Washington in 1960 but to Haloid Xerox. John F. Kennedy was elected president that year and Haloid Xerox introduced the 914 copier, the first office copier to use ordinary paper. Despite weighing six hundred pounds, the machine was wildly successful. Refinements like air nozzles to pull single sheets of blank paper into the machine without charging them with static electricity were developed. Americans added a new word to their vocabulary—toner—for the powdered ink that was developed in a Rochester garage.

An investment of over $60 million by Haloid Xerox had paid off and made millions for the company, which changed its name to the Xerox Corporation in 1961. Investors became millionaires, too.

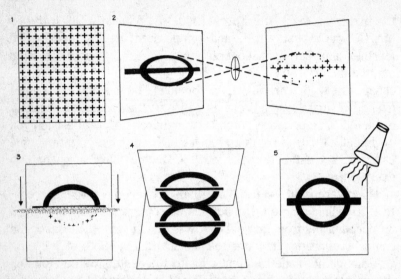

How Photocopying Works. A light-sensitive metal plate (1) or drum is positively charged. An image is projected onto the plate and is recorded as a pattern of positive charges (2). Negatively charged toner is sprinkled (3) on the plate and sticks to positive charges. Positively charged paper is placed on the plate, where it attracts the toner (4). Heat (5) fixes the toner to the plate.

Chester Carlson became immensely wealthy and proceeded over his lifetime to give away a hundred million dollars to charities and private citizens who wrote him, with most of his donations being anonymous. He died of a heart attack in 1968 in New York City.

Chester Carlson never became a household name, like Bell or Edison. But next to the telephone, what is the most widely used office machine? As for the "experts" who promoted carbon paper, they may be interested to know that the world's last remaining sizable carbon paper market remained in the Soviet Union, where photocopying had been practically illegal. Let's watch and see what the breakup of the Soviet Union does for the photocopier.[1]

12

....

Laser: The Atomic Radio-Light

I have heard articulate speech produced by sunlight. I have heard a ray of sun laugh, cough and sing.

> —ALEXANDER GRAHAM BELL,
> after sending words over a beam
> of light, 1880

WE OWE THE invention of the supermarket checkout scanner to a man sitting on a park bench, admiring the azaleas on a beautiful spring morning in Washington, D.C., in 1951. We are also indebted to this man for miraculous achievements in medicine, communications, engineering, surgery, weaponry, and other fields.

What Charles H. Townes did was to conceive the theory behind the laser, a highly concentrated beam of light that has influenced just about every product and service on earth.

Like many inventors, Townes did not set out to invent the laser. He had been racking his brains for a new way of producing very high frequency radio waves. These tiny waves would be able to probe deeply inside matter and allow minute pieces of it to be analyzed and measured. The molecular world of physics and chemistry would be opened.

But the vacuum-tube technology of the first half century forbade the generation of extremely small waves. There was a limit to just how small humans could make the "resonators" or cavities in which the tiny waves would be created. What next? That's when the park bench idea hit.

119

A MIDNIGHT STROLL

Townes had arrived in Washington to attend a conference sponsored by the Office of Naval Research. The goal was to study the
problems of producing these tiny waves, each less than one milli-
meter long. The problem racked his brain; he couldn't sleep. Thinking that a stroll in the park might free up some new ideas, Townes
dressed quietly so as not to disturb his brother-in-law, Arthur L.
Schawlow of Bell Labs, and walked around during the predawn
hours. His thoughts finally came to rest on a park bench in Franklin
Square. Then, out of nowhere, the idea struck.*

"That morning in the park, I realized that if man was to obtain
wavelengths shorter than those that could be produced by vacuum
tubes, he must use the ready-made small devices known as atoms
and molecules."[1]

Scribbling his ideas on the back of an envelope, Townes did not
discuss the new idea with committee members. He waited until he
got back to his laboratory at Columbia University to talk over the
concept and refine it. Basically, Townes was to draw upon a theory
for producing electromagnetic (radio) waves espoused by Albert
Einstein in 1917 called stimulated emission of radiation. According
to Einstein—and accepted by physicists in theory—given the right
initial conditions, if you drive radiation such as light or microwaves
past a group of atoms, they will be stimulated to give up energy.
Not just random energy but energy of exactly the same wavelength
(or frequency) and traveling in the same direction as the stimulating
beam. Townes realized that the energy that came out could be
reinforced and focused into an intense, steady beam.

---

* What Townes did not realize at the time of his nocturnal stroll was that his park bench was
located directly across the street from the former home of Alexander Graham Bell. In an
eerie twist of fate, Townes found himself sitting facing the home of a man who decades
before had invented the "photophone," a means of conveying pictures by light. The odd
coincidence was not lost on Townes. Later, when he learned who had lived there, Townes
said that thought sent a chill down his spine. For only after he sat down at the bench did the
idea of a laser beam come to him, only after he unknowingly stumbled across the residence of
the man who himself had shown that light could be used to carry information.

Later, after discussing the idea over lunch with his students at Columbia, Townes chose the name maser for such a device using microwaves as the energy source and the name laser to describe the process powered by light: *l*ight *a*mplification by *s*timulated *e*mission of *r*adiation.

In searching for a way of making tiny radio waves, Townes had hit the jackpot. Not only could his laser make short little waves but very powerful and incredibly straight ones. Lasers would revolutionize the world and earn its developers a fistful of Nobel Prizes.

THE WAVE

In theory, the idea behind the laser is very simple. Think of a crowd of people doing the wave at a baseball stadium. Here people stand up for a brief moment and wave their hands up and down in unison. The effect is the appearance of a wave of wiggling people traveling from one end of the stadium to another and back. To start the wave, just one or two people stand up. At this point, the wave is little more than a ripple as few people realize it's even passing. But each time the ripple travels back and forth across the stadium, a few more people stop eating their hot dogs and are stimulated by the excitement to join the crowd. The wave grows as more people join in. Eventually the crowd is a sea of people swaying and jumping in unison. What started out as a single, lonely person waving has, after many repetitions of the wave, turned into a powerful force flowing in one direction.

This is exactly what happens in a laser. Imagine that instead of a stadium that holds people we have a little ruby rod that holds atoms. Each end of the rod is polished into a mirrored surface. Let's say you start the wave by briefly shining a light onto the ruby. The light acts to excite a few of the atoms in the rod—like spectators at the stadium—and those atoms emit a light of their own. Some of the light travels parallel to the rod and excites a few more neighboring atoms, causing a little pulse of light to travel to the end of the rod where it meets the mirror. The light bounces off the mirror and travels back through the ruby atom "crowd."

On its way back, the light stimulates other atoms to give up light (radiate), and that energy is added to the wave of light. As the wave reaches the end of the ruby, it hits the mirrored end, and the process repeats itself. Eventually, after bouncing back and forth and stimulating enough ruby atoms to join in, the light beam has grown in intensity and the light waves reaching the mirror are powerful enough to burst through. A beam shoots out the end and, voilà, you have a laser. Because the light in the crowd stimulated identical atoms to join in, the emerging laser light is of one color and very intense.*

LIGHT FROM A RUBY

So much for theory, but could such an object be built? Townes would have to wait to find out. While he continued to have success building masers that amplified microwaves, no one could take the giant step of actually amplifying light itself. In 1958 Townes and Schawlow put their heads together and described such a laser in a technical journal and challenged anyone to build it. Laser mania broke out. Every physicist and technician with the available hardware set out to build one. The wait was not long; it ended in July 1960. Theodore H. Maiman, a physicist with Hughes Aircraft Company, demonstrated the first crude but workable ruby laser.

A ruby rod, polished to mirror quality at each end was surrounded by a glass curlicue flashbulb. Firing the flashbulb sent an intense beam of light out one end of the rod. Touting the achieve-

---

* As opposed to a regular light bulb, a laser puts out a pure color. Anyone who has looked at laser light sees the color as grainy. It appears as if the light beam is not solid. In reality, all the light is there. But it's so pure that not all the color-sensing cells in your retina—the cones— are stimulated to fire. Neighboring cones, naturally sensitive to multicolored sunlight, are left idle. We perceive the light as grainy.

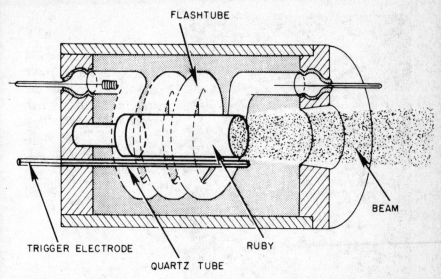

FLASHTUBE

TRIGGER ELECTRODE

QUARTZ TUBE

RUBY

BEAM

Elegantly Simple. The laser ruby emits a powerful laser light beam when the flash tube flickers.

ment as "an atomic radio-light brighter than the center of the sun," Maiman told the world that to generate a light wave as intense as that produced by a laser, a carbon-arc Hollywood klieg light would have to reach a temperature of several billions of degrees—a purely hypothetical example since the lamp would melt before that happened.

The laser was incredibly simple; it was elegant. And just about any kid with a junior high school aptitude and the right parts could make one. (I even tried to build one.)

Laser fever caught on with the public. Front pages of magazines showed lasers cutting through hardened steel razor blades.* The fabled "death rays" of comic books and movies had literally been invented. Truth *was* stranger than fiction. (Years later, in the

---

* A respectable laser was said to have at least one half "Gillette power." It could cut halfway through a double-edged Gillette Blue Blade razor blade. The Gillette was the accepted measure of power.

Atomic Radio-Light. The first laser was nothing more than a flash tube surrounding a synthetic ruby rod. Dr. Theodore H. Maiman, a pioneer in the development of lasers at Hughes Aircraft, is shown with his 1960 laser.

1980s, President Ronald Reagan would seize upon this extremely costly and difficult idea as the basis for his proposed Star Wars antimissile defense system.)

Ruby lasers were not very practical. They emitted one quick pulse at a time. Lasers made of gas soon followed. Excited by electricity instead of light flashes, gas lasers worked exactly like crystal lasers but could remain on indefinitely, producing continuous beams. Gas laser beams glow a lot like neon lights.

BOUNCING LASERS OFF THE MOON

It's the almost perfectly parallel beam that hardly spreads out when traveling that gives the laser a magical property. A flashlight, when shone across a room, will appear on the wall as a large dish-sized

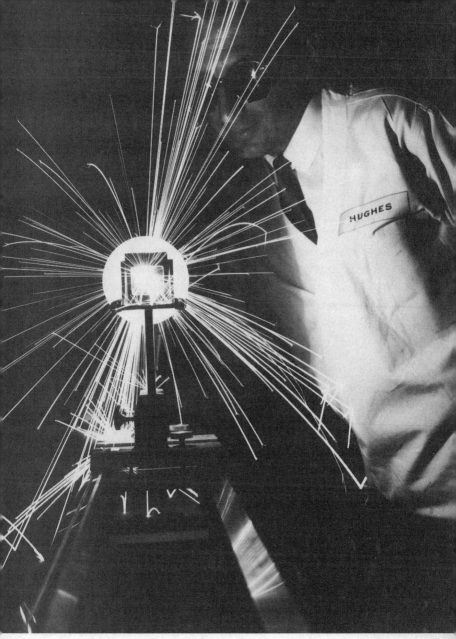

Don't Try This at Home. In its infancy, the laser was often impressively demonstrated by burning a hole through strong metal as in this 1964 photo of a laser slicing through titanium in less than one thousandth of a second.

circle of light. The rays have spread out as they've traveled. But a pencil-thin beam of laser light will still appear as thin. Laser light takes many more miles of travel to spread out than a flashlight beam. For example, when reaching the moon, nearly a quarter of a million miles away, a laser will light a lunar area just two miles across. If an ordinary searchlight could reach the moon, its beam would spread over twenty-five thousand miles. The distance to the moon has been measured within inches using laser beams bounced off reflectors left by *Apollo* astronauts.

A hundred watt light bulb is one hundred times as powerful as a one-watt laser. Yet that one-watt laser beam can readily be seen by television equipment set up on the moon. But not even all the lights on earth, shining from all the homes and office buildings at once at night, show even faintly on the moon.

When surveyors discovered this laser property they said, "Hey. What a great measuring device. A perfectly straight line. Let's throw away our sagging tape measures and use laser beams to measure distances and create straight lines." Not only did highway engineers rely upon laser beams for a true line but great underground tunnels and railroads that had to meet from work started many miles apart came within a fraction of an inch when guided by laser-based surveying equipment. They just aimed their drills at the glowing red dot on the tunnel wall.

WIRES OF GLASS

By the end of the 1960s more lasers were being designed and used by industry than by scientists for research. As the years went by, the original uses for lasers to cut holes and blow up enemy missiles were overshadowed by much more constructive and elegant tasks.

In the entertainment world lasers powered compact disks and provided magnificently colored light shows. Holograms, three-dimensional images produced by lasers, found their way to the covers of magazines and promotional giveaways.

Lasers proved perfect for surgery, for welding torn corneas and burning away blood clots. But the 1970s saw the dawning of the era of the laser's greatest strength: communications. Normal electrical wiring limits the amount of telephone, television, and computer traffic that can be carried by wire. But in theory, a single laser beam should be able to carry all the radio, TV, and telephone conversations of the entire world. Lasers pulsing at billions of flashes per second can already transmit tens of thousands of phone calls on optical fiber highways of glass. In 1977 the Bell System installed the first working laser cable system beneath the streets of downtown Chicago. It was the first to carry phone calls, computer data, and video signals on pulses of light.

The decade of the nineties is finding lasers being unleashed in computers. Scientists are now finding ways to hook millions of laser beams to computer chips. These "photonic switches" do away with the need of converting lasers into electrical pulses normally used by computers and allow the light beams themselves to be processed. A dramatic increase in computer power results.

FINALLY . . . LITTLE, TINY WAVES

Ironically, Townes's initial goal of creating very short electromagnetic waves has been dramatically achieved. The original ruby laser's light produced waves only twenty-seven millionths of an inch long. It's a technology hardly imagined forty years ago by a scientist sitting on a park bench. And it still astonishes today.

No walk of life has been untouched by laser magic. Making a sweeping generalization like that is dangerous. There's always someone who'll say, "I can name something the laser has nothing to do with." But in the case of the laser just about every product you can buy—from food to automobiles to books—every service you can think of—from consulting to dry cleaning—every profession imaginable—from doctor to ballplayer—can find a laser beam shining in its midst.

"It was by no means clear, even to those who worked on it,"

wrote Townes, "that it would see so many striking applications. And much undoubtedly lies ahead."[2]

BELL DID IT FIRST . . .

About four years after inventing the telephone, Alexander Graham Bell invented the first method of sending voice by light. Bell not only found a way to transmit speech by wire, this visionary designed a way of sending speech by light beam.

In 1880, seventy-seven years before the invention of the laser by the laboratory that carries his name, Bell patented his "photophone." The idea was to transmit speech via sunlight. But the photophone did not work very well. Sunlight can easily be scattered by air, smoke, and rain. And what does one do when the sun isn't shining?

What's in a Name?

The invention of the laser was an accomplishment eagerly sought after by the world's leading laboratories. But following its invention in 1958, the leading American physics journal refused to publish a paper reporting the triumph. Instead, in a comedy of errors, the paper wound up in, of all places, the British journal *Nature*.

Normally the home for research papers on frogs, cancer, and astronomy, *Nature* found itself welcoming the paper when the prestigious *Physical Review Letters* turned the announcement down sight unseen. They misconstrued the name. Maiman had entitled his submission "Optical Maser Action in Ruby." Optical maser was a common name for the laser in the late 1950s. Masers, the forerunners of lasers, were very popular back then. *Physical Review Letters* was overloaded with articles about the maser's wonderfulness and refused to speedily publish any new papers with the name maser in their titles.

According to AT&T, French scientist Ernest Mercadier suggested that since the photophone used radiant energy, it should be named the radiophone. In all probability that was the first time the word radio was used in connection with an invention.

SO WHO INVENTED THE LASER?

In the same time frame as Townes and using similar ideas, two Soviet scientists had been working to build masers and lasers. Because the cold-war politics of Soviet research had kept Soviet science out of the spotlight, it was impossible to tell whether the Soviet scientists, Drs. N. G. Basov and A. M. Prokhov, were ahead or behind Townes in their research or who did what first. So the Nobel Prize committee decided to award the 1964 Nobel Prize in Physics to all three for their work leading to the development of masers and lasers.

The prize may have awarded the "social invention" of the laser to Townes et al. A Nobel Prize carries great weight in the eyes of the public. But it didn't settle the "legal" invention argument—to whom to afford the patent rights. Another scientist, Dr. Gordon Gould, continues to insist that he share the legal if not the social credit for inventing the laser.

In 1957 Gould was a graduate student at Columbia University, the same university where Townes was teaching. Gould also saw that to amplify light it needed to be "pumped" back and forth in a confined space. He knew how important a breakthrough this laser idea would be. Some say he may even have been the first person to have used the word "laser." But Gould fell into a series of unfortunate events that to this day cloud the issue. First, instead of publishing his ideas in scientific journals, where they would be on record for everyone to see, Gould wrote them down in his notebooks, which he had notarized. Second, while he was farsighted enough to apply early for patents for his laser concepts, he naively gave up on the patenting process when given incorrect advice that in order to be granted a patent he needed a working model, which he did not possess.

The stage was set for a battle. In June of 1959, Townes and Arthur Schawlow took their laser idea and filed for a patent. Nine months later, Gould filed to patent his similar ideas. And a patent fight between the parties has been in the courts ever since, centered on who owns the rights to laser concepts. Gould achieved a limited victory in 1977 by winning two patents—one for laser pumping and the other for a method of processing materials with lasers. But the "laser wars" are continuing to this day with a tenacity that would make even Darth Vader blanch.

13

....

Velcro: Improving on Nature

Beauty in nature is a quality
which gives the human sense a chance
to be skillful.

> —BERTOLT BRECHT, *The*
> *Messingkauf Dialogues*, 1965

WHEN ASTRONAUTS AND cosmonauts first began circling the globe, they faced a unique problem: how to keep track of all their stuff. On the ground, if you wanted to stow your car keys, for example, you simply left them on the night table. They could be found in the morning right where you left them—perhaps after a bit of last-minute searching on the way out the door—but at least they stayed put.

Space travelers faced a different problem. Objects had a habit of just floating off. A pencil, a wrench, would just hang around drifting weightlessly wherever a tiny shove or air current would take it.

Keeping tabs on a capsule full of flotsam was a job in itself, a full-time job because if a tiny piece lodged itself into the hardware, it could short-circuit who-knows-what. The fact that orbital living quarters would make a phone booth look spacious only compounded the problem of keeping things neat and orderly.

How could one make sure things stayed put?

COCKLEBURS TO THE RESCUE

The answer came from another lucky walk in the country (see Chapter 23, "The Wasp That Changed the World") taken just a few years before the space race began in the early 1950s. This time our hero was one George de Mestral, who happened to be taking a stroll one day in his native Switzerland.

Upon arriving home, he found his jacket covered with cockleburs. Picking the sticky seed pods off his clothing, de Mestral wondered what act of natural engineering could account for their tenacious sticking ability. Whereas you or I might just curse the darned cockleburs for being such a nuisance, de Mestral pulled out

Cocklebur. George de Mestral's examination of a cocklebur, Mother Nature's original "push fastener," following a nature walk in Switzerland led to the development of Velcro.

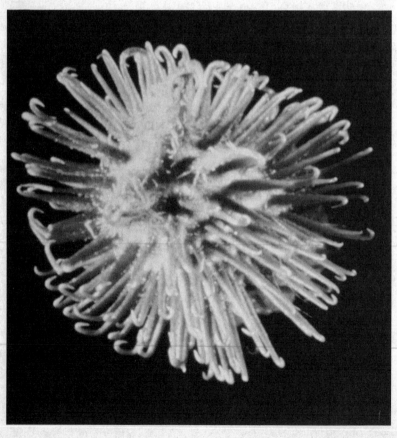

Velcro Under the Microscope. Thousands of tiny hooks engage countless loops. Velcro now comes in every shape and size—it's even made out of steel.

his microscope and took a careful look. Focusing in on the cockleburs' structure, he noticed they were covered with little hooks that entangled themselves in the loops of fabric of the jacket. Mother Nature had invented an ingenious method for catching a free ride to the next seeding spot by lodging her seed carriers in the fur of passing birds and animals.

If nature could be so resourceful, why not take advantage of her design and turn nuisance into necessity? Artificially create a system of hooks and loops that when pressed together tightly stick to one another but when pried apart easily separate. Velcro*—derived from *vel*vet and *cro*chet—was first made in France. Each Velcro tape was made by hand and took almost forever to produce. The loops could be easily made by machine but the hooks did not lend themselves easily to mechanization. What to do? Make the loops

* Velcro is a registered trademark for fasteners made by Velcro Companies.

mechanically and then cut them in such a way that the clipped ends formed hooks! This way hooks could be fashioned from loops from the simple act of cutting.

With the mechanical problems solved, Velcro's holding power was improved. The original nylon material used to make the hooks and loops was strengthened by thickening. Blends of polyester and nylon made them even stronger. NASA found ribbons of unique fastening material it needed to hold the countless odds and ends of space travel.*

Even today, no better substitute has been found. When getting ready to leave orbit, space shuttle astronauts literally spend a full day in space collecting all of the material Velcroed to the walls.†

Of course Velcro has been improved over the years. It has become impervious to water, chemicals, and corrosive ultraviolet light. Extra-strong Velcro can be made out of stainless steel and synthetic fibers that withstand temperatures upwards of 800°F. and do not catch fire.

---

* About ten to fifteen pounds per square inch of force is required to pull apart standard Velcro. A ton of force is needed to pull apart two hundred square inches.

† President Bush's White House staff has coined the word velcrosis to mean a situation in which people uncontrollably tend to swarm around and "stick" to the President when he makes a public appearance. You can watch velcrosis in action the next time any famous person is greeted by a crowd of celebrity-seekers.

14

Teflon: The Top-Secret Discovery

The lowest coefficient of static and dynamic friction of any solid.
—*Guinness Book of World Records*, describing Teflon

Roy J. Plunkett didn't know he had invented Teflon. Teflon appeared totally by surprise to the young chemist. Just two years earlier he had graduated with his Ph.D. from Ohio State University. And now, on April 6, 1938, as a chemist for Du Pont, Plunkett had unknowingly invented a compound that would change the world.

Plunkett had been assigned a project: Come up with a nontoxic refrigerant. Jack Rebok, Plunkett's lab assistant at Du Pont's Jackson Laboratory in New Jersey, had just cracked the valve on a bottle of special Freon gas that Plunkett had concocted. But no gas came whistling out. Scratching his head, Rebok turned to Plunkett.

"Hey doc, did you use all this stuff up last night?"

"No. I don't think so," replied a perplexed Plunkett.

"Well, there's nothing coming out," said Rebok as he examined the opened valve.

"Well that's odd. Let's check it out."

Trying to understand where the gas had gone, Plunkett suggested they weigh the cylinder. "It weighed what we expected it to, so we knew something was in there."

Perhaps the valve was stuck? Running a wire through the valve showed it to be open. The only thing left to do was open her up. Sawing open the cylinder, Plunkett dumped out its contents. "I was flabbergasted," he said. "Gee whiz, it's gone wrong."

And sure enough, out of the innards of the cylinder came not a gas but instead a greasy white powder. What was this stuff?

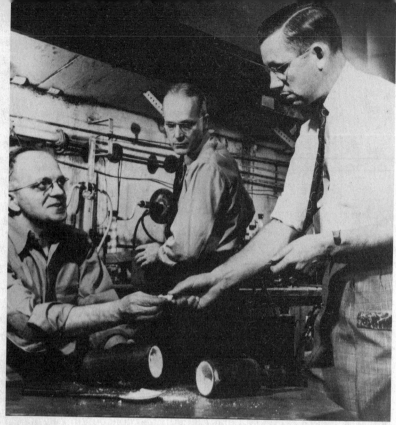

Serendipity in Action. Dr. Roy Plunkett (right) reenacts the discovery of Teflon in his lab. Plunkett shows his assistant, Jack Rebok (left), the white powdery material he scraped from inside the cylinder they curiously sawed open after a "failed" experiment with refrigeration gases. Bob McHarness looks on during this reenactment of the 1938 discovery.

SERENDIPITY IN A BOTTLE

Plunkett had been looking for a new kind of refrigerant, the gas used in air conditioners and refrigerators to suck out the heat. He thought he was mixing together a batch of tetrafluoroethylene—Freon—a little-used substance that might have refrigerating potential. But now, instead of finding the gas in his bottle as expected, he came face to face with a mysterious white powder.

If Plunkett had been a highly focused inventor like Edison with his light bulb, he might have just tossed the stuff out and written the whole thing off as "mysterious dirt in collecting cylinder" as Edison had written of the Edison effect. "But that's the thing about a discovery," Plunkett would say later, "even though it [the experiment] did go wrong you go ahead and find out what you got." Being the curious scientist that he was, Plunkett gave it the usual preliminary tests: Rub it with your finger, taste it, sniff it, see if it burns, drop some acid on it.

Nonstick Pages. Two pages from the 1938 laboratory notebook of Roy Plunkett, ushering in the age of nonstick coatings. Based on the work begun on these pages, Plunkett, holder of the original patent for Teflon, was invited to join Thomas Edison and the Wright Brothers in the National Inventors Hall of Fame.

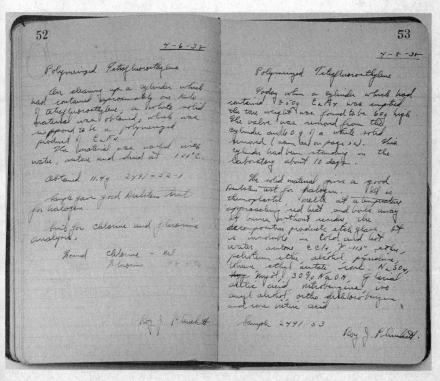

What neither Plunkett nor anyone else realized on that day, April 6, 1938, was that he had accidentally invented Teflon,* a material so revolutionary and valuable that its discovery would remain a military secret until after World War II. Inside the tank the molecules of the gas had formed into a long chain (polymerized) so that they now became a solid. More precisely, a plastic. One of the first ever invented.†

Plunkett soon discovered that Teflon was truly unique. It was an inert substance. It was absolutely stable; nothing reacted with it. Neither heat nor electricity nor acids nor solvents could affect it. It was corrosion-proof. And to top it all, Teflon was the most slippery substance on earth. More slippery than the slipperiest substance in nature: wet ice against wet ice.

A MILITARY SECRET

A few months later, Teflon was put to the test. Scientists working on the first nuclear bomb—the Manhattan Project—needed to make gaskets that could resist the terribly corrosive uranium hexaflouride used in making U-235, the chief ingredient in the bomb. They turned to Du Pont.

When told that the experimental new plastic was very expensive, General Leslie R. Groves, the army's ranking officer on the project, said that money was no object. Secretly, Du Pont made more of the white powder during the war and supplied it to the military. While Teflon and nylon were both invented in the same year, Teflon was not known to the public until after the war, 1946.

One of the first postwar customers was a gasket company in

* The name Teflon is derived by combining the chemists' nickname for tetrafluoroethylene—*tef*—with an arbitrary suffix—*lon*—that Du Pont likes to choose for its products, as in nylon, Orlon, and so forth.

† Plunkett and Du Pont refer to Teflon as being "discovered." To be precise, Teflon *was* discovered when the bottle was opened. But Teflon does not occur in nature so I prefer to call it an invention. The National Inventors Hall of Fame agrees: It inducted Plunkett into its hallowed halls in 1985.

Cameron, N.J., that bought the powder from Du Pont in 250-pound kegs. The company had borrowed money from a bank. To protect its interest, the bank required the gasket company to store the valuable stuff in its bank vaults overnight. The company would buy the Teflon and the Cameron bank took it as collateral, allowing the gasket company to dip into the stock in the vault when needed.

SOLD OUT AT MACY'S

The secrecy shrouding Teflon was lifted after World War II. But Teflon wasn't commercially available until 1948 and didn't become widely used until the late 1950s. Du Pont scientists needed many years to understand its chemical properties and to work out methods to produce it and then fabricate it into coatings, wire insulation, and component parts.

Tin Man. Roy Plunkett showing off cable insulated with Teflon and muffin tin coated with the nonstick surface. Muffin tins were among the first items of cookware coated with Teflon. Teflon was found to be the most slippery substance ever made, so slippery that the nonstick coating refused to stay bonded to early cookware. It could be easily scratched off with a metal spatula.

Bakeries were the first to use Teflon-coated muffin pans. As for the public, the first frying pan coated with Teflon was actually an import from France. The French introduced the pan in May 1956 at La Riviera, a store in Nice. Teflon-coated pots and pans sprang up on Macy's shelves just in time for Christmas 1960. Highly priced at $6.94, they sold out despite the fact that over seventeen inches of snow had blanketed the city two days before the sale and the temperature was a bracing nine degrees.

Those early Teflon pans were not without problems. Though the pans were touted for their nonstick surfaces, consumers quickly found out how nonstickable they really were. The Teflon was so slippery that it wouldn't stick to the metal pans. A spatula coraling a sunny-side-up would easily scratch the Teflon off. Scouring pads took chunks away. The public became disillusioned with Teflon. Du Pont improved the bonding by imbedding the Teflon in the metal, calling its product Silverstone.

In the mid 1970s the public was turned on to Teflon clothing. A light, warm, breathable fabric made of Teflon was invented. It was quickly pounced upon by outdoors folks for their cold-weather camping and skiing wardrobes. Its name? Ten points if you said Gore-Tex (see the section on Gore-Tex later in this chapter).

Despite Teflon's boon to the consumer, its uses in cooking and clothing are at best trivial. It's like saying penicillin's best use is to treat a minor cut. There is no person on earth whose life has not in some way been affected by Teflon. Teflon is found in bridges, boats, and planes. The George Washington Bridge linking New York and New Jersey has sheets of Teflon filled with graphite installed as bearing pads.

ICON OF THE SPACE AGE

Teflon is the original space-age material. John Glenn's spacesuit was coated with Teflon. Teflon has gone to the moon, has passed Mars, Jupiter, and the solar system. It can be found in nose cones, heat shields, and fuel tanks of spaceships, space probes, and satellites.

Its widest use is in the electronics industry. Miles of telephone and computer cable lining office buildings are insulated with the stuff. There are Teflon parts in televisions and pacemakers. Millions of people are functioning with artificial arteries and veins and replacement knee joints made of Teflon. Heart patches and improved surgical thread are made from Gore-Tex. The body does not reject Teflon.

Because it resists corrosion and is a great insulator, the best wires and cable are coated with it. It is ubiquitous in the electronics industry. If you've ever wondered how a bridge expands and contracts in the heat and cold, Teflon plays a large part. The rollers the bridge rides are made of Teflon.

Because it's flame retardant and resists weathering, Teflon is used to coat the ten-acre fiberglass fabric dome roof of the Silverdome in Pontiac, Mich. Not to be outdone, the Saudis have covered the King Addulaziz International Airport in Saudi Arabia in a fiberglass/Teflon roof.

Even the Statue of Liberty owes its future longevity to the slippery stuff. Teflon coatings and spacers serve to insulate the copper skin from the inner stainless steel framework, thus preventing corrosion of Lady Liberty.

As for Dr. Roy J. Plunkett, after thirty-nine years of dedicated

---

What Makes Teflon Tick?

Teflon is a polymer—a long stringy molecule made up of a chain of carbon atoms, the same as all polymers (which are the basic building blocks of plastics). What makes Teflon unique is that its carbon chain is completely surrounded by fluorine atoms. The bond between the carbon and fluorine is very strong; and the fluorine atoms shield the vulnerable carbon chain. This unusual structure gives Teflon its valuable properties. It's extremely slippery, inert to almost every known chemical, has excellent electrical characteristics, and remains useful under extreme temperatures: from $-400°$ F to $500°$ F.

work for Du Pont, he retired and was inducted into the National Inventors Hall of Fame in 1985. And while granted a patent for Teflon, Plunkett never received any royalties. Just a bonus and the world's gratitude for not discarding the greasy white powder.

THE SEARCH FOR A NEW REFRIGERANT

History is repeating itself. Ever since scientists discovered that fluorocarbons like Freon may be depleting the protective layer of ozone that shields the earth from harmful ultraviolet rays, they have been racing to discover a safe substitute. When Roy Plunkett was assigned the job of finding a safe refrigerant, the world was looking to replace the toxic and harmful refrigerants of the day: ammonia and sulfur dioxide. Plunkett turned to fluorine compounds because of a serendipitous incident that had occurred ten years before, in 1928.

Two General Motors Frigidaire Division scientists, Thomas Midgley, Jr., and Albert Henne, had scoured the scientific journals and decided that the ideal refrigerant would be carbon-based and contain both fluorine and chlorine. These compounds were known not to be toxic. But there were reports of some cases in which the fluorine compounds were poisonous.

What to do? Why not make some and test their toxicity on animals? Easier said than done. The total U.S. supply of raw ingredients needed to make a simple fluorocarbon consisted of five bottles of antimony trifluoride. No problem. Just order all of the bottles from the supply house. Which is what they did.

Choosing one of the five bottles at random, they prepared a sample and placed it next to a guinea pig to see what happened. The animal was unharmed by the gas. It appeared that the fluorine compound was nontoxic, as expected.

Being the good scientists that they were, the two decided that just to be sure of their results, they'd better perform the test a few more times. That's when the unexpected occurred. Preparing samples using the other four bottles, the scientists found that every guinea pig they tested died immediately from the vapors! Was the first test a fluke? Examining the contents of the bottles, the researchers found that each of the last four had been contaminated with plain water. But when mixed in with the antimony trifluoride, the water led to the formation of deadly phosgene gas.

A HAPPY ACCIDENT

Had Midgley and Henne chosen any of the other four contaminated bottles first, they might have incorrectly assumed that fluorocarbons are indeed deadly. A whole family of compounds, which would later go one to become the refrigerant of choice, might have been eliminated. Teflon would probably never have been discovered. The world would be a different place today.

Instead, due to a fortuitous twist of fate, Du Pont and Frigidaire went on to join forces, leading to the production of Freon (the Du Pont trade name for fluorocarbons) and setting the stage for the appearance of Roy J. Plunkett and his magical Teflon ten years later.

FROM TEFLON CLOTHING TO BODY PARTS

Of the twenty men who worked on the Teflon research team at Du Pont, Wilbert L. Gore could not get the product out of his system—in the figurative sense, of course. Bill Gore kept seeing new products that could be fabricated from this miraculous molecule. For example, Gore found a way of fashioning Teflon ribbons into electrical wiring insulation. So when Du Pont disbanded the Teflon team in 1957 and sent the scientists on to other projects, Gore pleaded with the company to allow him to develop new Teflon products.

They wouldn't hear of it. Gore did what any self-respecting techno-junkie would do: He went down to his basement and experimented on his own time. Taking a sample of the Teflon polymer PTFE (polytetrafluoroethylene), Gore continued working on his electrical insulation scheme and succeeded in making a unique product: flexible computer wiring cables with the magical insulating properties of Teflon. No other company had such a product to offer.

Surely now Du Pont would have to take notice. So Gore once again approached Du Pont and again was turned down. All right, he said, if you can't see the merits of this stuff, I'll go it alone. At the age of forty-five he left Du Pont and went into the Teflon wiring business.

BORN IN THE BASEMENT

Gore spent the next two years building computer cables out of PTFE bought from Du Pont. His business was completely owned and operated by his family. He ran the business from his basement.

The 1960s were the beginning of the transistorized electronics era, and here was a man with the right product at the right time. Teflon coatings filled the growing demand of electrical wiring. Gore made his business immensely successful. He moved out of his basement and eventually opened up two factories to make Teflon cabling.

His oldest son, Bob, was drawn to the family business. Holding a Ph.D. in chemical engineering, Bob had his father's childlike curiosity. He was ready to take Teflon one step further, just as his father had.

Bob and Wilbert reasoned that if PTFE could be fashioned into cable ribbons, why not into fabric? They realized that if the layers of the PTFE molecule could be unfolded, the resulting fabric would become "breathable"—the fabric would contain billions of microscopic holes per square inch. These pores would be large enough to allow air molecules to pass through but small enough to prevent water molecules from entering.

To create this special Swiss cheese, the Gores had to stretch the fabric extremely thin. Really thin. Like stretching pastry dough. But every time they heated the plastic and slowly stretched it, the darned thing snapped before it would stretch thin enough.

This was truly frustrating work. After several unsuccessful tries,

Bob Gore got sore. Enough of this elegant trial-and-error method. In a pique of anger, he wrenched a rod of the stuff from the oven and gave it a sudden, violent tug. And, voilà—in one serendipitous event—instead of snapping like taffy, the short rod stretched the length of his arms.

Now the Gores had a raw material they could make into fabric, and that's exactly what they did. But what? What kind of object would best be made out of a fabric laced with billions of tiny little pores? Something that needs to "breathe."

A tent seemed the natural answer. Anyone who's spent a night camping out knows that come morning your sleeping bag and clothing—not to mention your skin—are moist and clammy due to the water vapor, dew, and rain trapped inside the tent overnight. Tent designers had fashioned ingenious methods for venting tents to eliminate the problem. But suddenly the Gores had hit upon an idea no one had been able to implement: Make the tent itself breathable.

Fashioning a tent out of their newly named product, Gore-Tex,* Bill and his wife took the tent on an overnight field test. On their first night out a heavy rain pelted the tent. Here was a true test of the material's breathability. Feeling around the inside of the tent for leaks and finding none, the Gores were elated. Their joy was short-lived. A hailstorm followed and tore holes in the tent, filling it with water. They did not sleep well that night.

Back at the drawing board, the Gores decided they would have to strengthen the fabric. But simply adding more layers or thicker material was not the answer. For Gore-Tex to be successful in a wide variety of products it had to appeal as a comfortable piece of apparel, and at this point it suffered from major drawbacks. Besides being so bulky that it literally stood up by itself, this first Gore-Tex fabric would degrade with age. The super-thin membrane would weaken and allow body moisture—sweat—to penetrate when it came in contact with body oils. The problem became so acute that a whole line of Gore-Tex had to be recalled.

* Gore-Tex is a trademark of W. L. Gore & Associates, Inc.

Undeterred, Bob Gore redesigned his membrane. He shifted the molecules around, making them impervious to body oils. He perfected a method of laminating the membrane onto soft fabrics to give it body and strength. Then he watched sales of Gore-Tex skyrocket.

TEFLON BODY PARTS

If the Gore-Tex story had ended there, it would have been a fairy tale of the fashion industry. But lost in the glare of success in sporting goods is one of Gore-Tex's greatest but quietest achievements. It started back in 1971, ten years before Gore-Tex was just coming into its own in the sportswear industry.

The Gores had been taking one of their regular ski trips to Vail, Colo. A member of this vacation trip was a physician from Denver General Hospital, Dr. Ben Eiseman.

During a break, Bob yanked a piece of Teflon tubing out of his pocket and showed it to the doctor. When told of the inert properties of Teflon, Eiseman offered to implant the tube in a pig and see what happened. Doctors were always looking for new materials that would not be rejected by the body.

A few weeks later, an excited Eiseman phoned Gore to tell him that his experiment had been a success—the pig did not reject his Teflon tube. That was the beginning of the Teflon age of medicine. Since then, millions of people have had artificial body parts made out of Gore-Tex—arteries and ligaments—successfully implanted into their bodies. The pores make an ideal place for the natural tissues to grow into, and the inert properties of Teflon fool the body into thinking Gore-Tex is not a foreign object.*

People can wear Gore-Tex on both the inside and the outside of their bodies at the same time.

---

* Roy Plunkett, discoverer of Teflon, tells about the time "I was at a dance and a doctor friend asked me if I wanted to meet somebody. He said 'This guy is here only because he has a Teflon aorta which I installed.' "

15

....

Nylon: The Hit of the '39 World's Fair

We Enter the World of Tomorrow
—New York World's Fair, Forum
 on Current Problems, 1939

THE 1939 NEW York World's Fair will be remembered for many things: for Americans' first glimpse of television, for their first "rides" in space.

But to three thousand excited members of a women's club, the World's Fair site was their first keyhole to the future of hosiery. For at Flushing Meadow that October day in 1938, where the TV sets of tomorrow, the rides into the future, and the outdoor theater would sit a year later, E.I. du Pont de Nemours, Inc., chose to show these women a material never seen before on earth. Unlike wool, cotton, or silk, which were made out of naturally occurring fibers, this one, in the words of Du Pont vice president Charles Stine, was "the first man-made organic textile fiber wholly from new materials from the mineral kingdom."

Du Pont called it nylon.

THE AGE OF THE MAN-MADE FIBER

As a product of science, nylon carried almost mythical properties. "Though wholly fabricated from such common raw materials as coal, water, and air," Stine told the silenced audience, "nylon can be fashioned into filaments as strong as steel, as fine as a spider's web, yet more elastic than any of the common natural fibers."

What more could any clothes-conscious woman ask for? They erupted into applause. They were sure they had just witnessed the dawn of a new age: the indestructible stocking. No longer would unsightly rips and tears embarrass these ladies. No longer would the edge of a file cabinet drawer be a lethal weapon. The run was dead. Before long women would wait hours in line to purchase their "nylons."

Nylon would change the course of fibers and fabrics around the world. It heralded the era of "man-made" fibers with trade names like Dacron and Orlon. But none of the smiling women in the audience knew that the product that would become the biggest cash cow in the history of Du Pont was the result of a lucky accident back in the labs.

Wallace Hume Carothers had come to the Du Pont labs in Delaware in 1928 to conduct basic research—that is, he really didn't know what direction his work would take, where it would end up, or if it would produce anything at all. The idea behind basic research is to find out as much about something as you can, and if that path leads to something useful, all the better. Some of the world's greatest discoveries and advances had started this way. Carothers had been spirited away from Harvard University and been guaranteed by Du Pont that he would be allowed to continue doing his pure science. Being a brilliant researcher, Carothers was given enough line by the company and allowed to run.

SPAGHETTI MOLECULES

Du Pont's Stine told the young idealist Carothers that the direction the company wanted him to take was in opening up the emerging world of polymers: long, stringy, spaghetti-like molecules that gave rubber and silk their special properties. By learning how they were constructed, it might be possible to create man-made fibers out of raw materials found cheaply in the mineral kingdom. After all, if a silk worm could do it, why not a human?

What scientists around the world didn't know at the time was what held those spaghetti molecules together. What were the forces involved? Were they the same as those that held smaller molecules together, or were they unique?

One way to find out, thought Carothers, was to build those molecules from scratch—synthetically—and to find out in the process what binds them. By 1930 Carothers and his colleagues had reached the conclusion that polymers were molecules just like all others—only longer—and held together the same way. By April of that year, using this knowledge, they had accidentally made the first synthetic rubber, neoprene. But then their work hit a dead end.

THE ACCIDENT THAT CHANGED HISTORY

By the middle of 1933 the fiber research had come to a halt. The research team had learned a lot about polymers, but they had not hit the jackpot. Carothers decided in early 1934 that he was a failure; there would be no synthetic version of silk. Then came the kind of accident that makes inventions famous.

Carothers had created a new substance—a polyamide—with a structure similar to silk but without silk's properties. Its melting point was too high to be spun into fibers, so what good was it? Carothers stuck his product—nylon—on the shelf without even bothering to patent it. His colleagues at Du Pont turned their attention to a more useful class of plastics, polyesters, which were easier to work with in the laboratory. One of those intrigued by the properties of polyesters was Julian Hill. While fooling around in the lab with a beaker full of the stuff, Hill noticed that when he stuck a glass stirring rod into the polyester, fixed a glob of it onto the rod, and drew the rod out, the polymer would stretch into silky strands as thin as a spider's webbing.

This curious property of "cold-drawing" intrigued the entire lab. And as the story goes, they all waited until the boss (Carothers) was off the premises and conducted an unorthodox experiment of

Like Pulling Taffy. Dr. Julian Hill reenacting the discovery of nylon.

their own. Affixing another ball of the gooey plastic to the end of a rod, Hill and his lab buddies decided to see how far they could stretch it. Much to their surprise—and delight—they could stretch the polyester string the entire length of the hallway.

The act of cold-drawing and stretching actually increased the strength of the material by aligning the molecules. And if it worked for these polyesters that had melting points too low to be made into fabric, what about that old, useless polyamide stuff that Carothers had shelved?

Sure enough, it worked. Cold-drawing was the secret to creating nylon. It gave nylon excellent qualities; nylon was stronger and stood up to abuse better than silk.

Within five years Du Pont brought nylon onto the market. Nylon was aimed squarely at the hosiery-buying public. Du Pont resisted efforts at making nylon a substitute for such commodities as leather, wool, and cellophane, and in this way could concentrate on keeping the price below the cost of silk. Of course, it also helps to have a monopoly on a product.

EPILOGUE

Wallace Hume Carothers never lived to reap the fame and fortune of his discoveries. Prone to depression, Carothers suffered various bouts over a period of two years. In the summer of 1936 he succumbed to a nervous breakdown. The sudden death of his sister did not help matters. On April 29, 1937, Carothers took a fatal dose of cyanide and committed suicide in a Philadelphia hotel room. His death came just three weeks after the patent for nylon had been filed and two days after his forty-first birthday.[1]

16

....

Vaseline: Wonder Jelly from Oil Fields

The true scientist never loses the faculty of amazement.
—HANS SELYE, *Newsweek*, 1958

IN THE DAYS before big-budgeted laboratories and big-ticket science, tinkerers and inventors had no place to turn to test out their new ideas except themselves. This is a story of one such researcher, a young chemist who embarked on a campaign of inflicting minor cuts and burns on himself to see if his magical new salve could heal the wounds. It's the story of Robert A. Chesebrough and his wonder jelly called Vaseline.

This story goes back to the summer of 1859, when America's love affair with petroleum began with the discovery of huge oil deposits in the fields of Titusville, Pa. Even before the days when automobiles gulped a refined petroleum product called gasoline, the oil lamps of the nation sipped illuminating oil, also refined from crude oil.

Here we find Robert Chesebrough (1837–1933) all of twenty-two years of age and self-employed in the illuminating oil business. Born in England to American parents and then living in Brooklyn, N.Y., Chesebrough decided it was time to make a pilgrimage to the oilfields and strike up a deal with the oil barons.

MIRACULOUS CRUDE OIL

Chesebrough is also intrigued by oilfield folk stories attributing magical properties to the oil coming out of the ground. Oil workers tell of a miraculous healing substance, a residue that is left behind by the oil.

Withdrawing his life savings, Chesebrough buys a ticket to Titusville and makes his way to the noisy, black Allegheny oilfields. Quietly observing the scenery, he notices the sweaty oil workers and their amazingly dirty hands. Homing in on an operating oil pump, he observes a man scraping a waxy goo from the machinery. Being a friendly sort of guy, Chesebrough asks the laborer what the stuff is.

"Rod wax," is the answer.

"What's it do?"

"Nothing. Comes from the oil. Just coats the pump rods. Fouls up the works unless we get rid of it. Got to scrape it off."

"Pretty awful, huh?"

"Worthless. But tell you what. When the fellas burn or cut themselves, a little bit of this on the wound fixes it right up."

Walking away in a cloud of thought, Chesebrough wonders what magical healing ingredients could be locked up in the oil. As dollar signs drift before his eyes, he wonders if he can find a way of refining and marketing the magic grease. One could make a killing in rod wax.

THE BIRTH OF PETROLEUM JELLY

Grabbing a bucket, Chesebrough loads up on the black wax and takes a sample back to his Brooklyn laboratory. Day and night he tries to make some useful medication out of the black gold.

Our clock turns ahead ten years. By now, after a decade of experimenting with the Pennsylvania crude, Chesebrough has found a better product than rod wax. By distilling the lighter, thinner oil products from the crude, he has created a light-colored gel. After further purification his nouveau jelly takes on very desirable properties: It doesn't spoil the way lard, tallow, or vegetable oils do; instead it's pleasantly odorless.

Perhaps he's latched on to something truly marketable here. For years people have been smearing smelly goose greases, heated olive oils, and stinking garlic oils on their bodies to ward off colds and diseases or to relieve cuts and burns. Wouldn't people just love to have a soothing, odorless product that works just as well?

But first, Chesebrough has to be sure his miracle jelly will work. Conventional wisdom says he should try it out on something or someone. He needs an unlimited supply of fresh cuts and burns; all he has is his own body. He'll have to serve as his own guinea pig.

HARD-HAT GUINEA PIGS

So Chesebrough sets out to mutilate his own body. Inflicting scores of cuts and burns on himself with blades, flames, and acids, he proceeds to conduct his own crude consumer product campaign. And this isn't just an overnight fling. Chesebrough abuses himself for weeks with all sorts of physical torture, but is always careful to apply a dollop of soothing salve to the wound.

And when he runs out of patience, he sends for more patients. Positioning himself at construction sites, he waits for workers to suffer on-the-job injuries. And when they hurt themselves, to the rescue comes Brooklyn's answer to Florence Nightingale with a soothing magical medical gook. And it works. Chesebrough finds that both his bruises and those of the construction workers heal better and hurt less under the spell of the mysterious grease.

Hearing opportunity knock, Chesebrough opens up a factory in Brooklyn in 1870. But no one breaks down his doors to buy even a single bottle. Though he gives out samples to doctors and druggists, they never call him back to reorder. Undaunted, Chesebrough realizes he has to create a market for his product. So he loads a wagon full of one-ounce bottles, hitches up a horse, and drives around New York state giving out free samples to anyone he can find. Here is the original medicine man, selling—no, giving away—his magical potion out of the back of a horse-drawn wagon.* And again, it works.

* Could this have been the birth of the free-sample promotion?

When people run out of the ointment they go to their druggists, who, of course, have none. Soon Chesebrough begins receiving orders from pharmacists around the state. Expanding his business to other states, Chesebrough soon has a dozen wagons crisscrossing the countryside, showering passersby in a deluge of free salve. Business could not be better.

FROM ROD WAX TO VASELINE

All new products need a catchy name. What to call this petroleum jelly? Chesebrough takes the Greek word for "oil," *elain*, and sticks it on to the end of the German word for "water," *wasser*. The result is the word Vaseline.

By the mid 1870s, Vaseline has crossed the Atlantic. Offices are opened in London and later in Spain and France. Vaseline has become so popular that in 1876 the prestigious British medical journal *The Lancet* gives it an unequivocable endorsement:

This new remedial agent is offered as a basis for ointments and emollient application for the skin, and is an American preparation of petroleum from which the lighter portions have been separated by distillation, and all taste and smell removed. It is perfectly neutral, does not become rancid on exposure to the air, will not saponify, and, in fact, is only altered with great difficulty by chemical agents. It is very soft, and altogether seems admirably adapted to the purposes for which it is intended. We think it will be very valuable in medicinal practice, and advise its careful trial.

Vaseline makes Chesebrough very wealthy. Approaching death at the age of ninety-six in 1933, Chesebrough cheerfully attributes his longevity to his daily use of his petroleum distillate.

17

....

Synthetic Sweeteners: All by Accident

Some of the powder got onto my fingers. When licking my finger to pick up a piece of paper, I noticed a very strong, sweet taste.

—JAMES M. SCHLATTER,
discovering aspartame, 1965

WHEN CONSTANTINE FAHLBERG sat down to dinner one night in 1879, he found that his bread tasted unnaturally sweet. Bread at that time was usually baked at home. So it's normal to assume that Fahlberg asked his wife about the recipe.

"Honey, did you add something sweet to the bread dough?"

"No dear. Tough day at the office?"

No written record of such a conversation has been found, so we'll never know.

What we do know is that Fahlberg, a research fellow at Johns Hopkins University, didn't merely assume some kitchen accident had sweetened the dough. Instead he surmised that something sweet from his lab must have stuck to his hands and ended up in his mouth at dinnertime. And the only way to find out was to taste everything he could in his lab (this is *not* how chemists sample chemicals anymore).

FINGER-LICKIN' GOOD

And that's exactly what he did. Sampling bottle after bottle of chemicals like a bloodhound with a sweet tooth, Fahlberg found the

sweetness he was looking for: a chemical called benzoyl *o*-sul-
fonamide. Possibly with an idea toward marketing the chemical,
Fahlberg wisely called the substance saccharin, after the Latin
word for "sugar," *saccharum.*

Calling saccharin sweet was like calling the universe large. Saccharin turned out to be the sweetest substance on earth: three
hundred to five hundred times sweeter than sugar.* Realizing the
immense wealth that could be made from his discovery, he received a patent for a commercial version of saccharin in 1885.

But the sweet smell of success left an acrid taste. When he
discovered saccharin, Fahlberg had been working in the laboratory
of and under the supervision of Ira Remsen, one of the most
respected and brilliant chemists of the time. Remsen brought
Fahlberg to Baltimore from Leipzig, Germany, after Fahlberg had
written Remsen and asked to be Remsen's student at Johns Hopkins, and set him to work on a project. It was during this work for
Remsen that Fahlberg discovered the sweetness of saccharin.

Instead of sharing the credit of the discovery with his lab boss as
a team normally does, Fahlberg claimed sole credit and secretly
obtained a U.S. patent. He awarded half-interest to a German
company who manufactured and marketed the artificial sweetener.

Fahlberg made a mint. Remsen didn't get a nickel. And he didn't
want one. All he wanted was "a little credit for the discovery."
After all, he said, they had published the account of its discovery in
a research paper authored by both of them. Remsen could have
contested the patent—the Merck pharmaceutical company was
ready to pick up legal fees—but Remsen declined.

If not the fortune, Remsen got the fame. Accounts of his fight
with Fahlberg increased his prestige in the research community,
which bestowed numerous honors and awards on Remsen. Ira
Remsen's other research made his co-discovery of saccharin seem
small. He is best remembered for his talents as an educator,
credited with establishing graduate research in chemistry in the
United States.

---

* Believe it or not, saccharin is also antiseptic.

TEDDY ROOSEVELT TO THE RESCUE

The last chapter of the Remsen-saccharin saga has an ironic ending. Saccharin became involved in a crusade by Harvey W. Wiley, of the Department of Agriculture, to rid the public of unsafe foods and drugs. This was at a time before the Food and Drug Administration and its requirements for safety and efficacy. Wiley, a tough and savvy politico, was able to shoehorn the first pure food and drug laws through Congress. Designed to protect the public from contaminated or adulterated food and harmful additives, the bill was signed into law by President Theodore Roosevelt over the howls of the food industry.

Flushed with success, Wiley set his sites on saccharin. Declaring the sweetener "a coaltar product totally devoid of food value and extremely injurious to health,"[1] Wiley sought to have it banned. Unfortunately, the President disagreed. Consuming saccharin daily on the advice of his doctors, Roosevelt could see nothing wrong with the substance. At his bully best, the President announced that "anyone who said saccharin was injurious to health is an idiot." To clip the wings of the upstart Wiley, Roosevelt appointed a review board of scientists to study each case Wiley wished to prosecute under the new food and drug law. Installed as the head of the review board was none other than Ira Remsen. Needless to say, the board found saccharin to be safe.

It wasn't until 1970, after extensive testing of saccharin on laboratory mice had shown an increase in bladder cancer, that the safety of saccharin came under serious attack. But by then, it had been joined by another class of artificial sweetener whose discovery was just as accidental—cyclamates.

REMEMBER SUCARYL?

The sweetness of cyclamates was discovered in 1937 at the University of Illinois by a chemistry graduate student having a problem with his smoking habit. Michael Sveda was working to discover a new drug that would kill bacteria as effectively as did the antibiotics of the day, sulfa drugs.

A confirmed pipe smoker, Sveda was having trouble getting used to his new smoking affliction: cigarettes. The chemistry student hated cigarettes; he wanted no part of them. But he had no choice. A year before a flood had swept through the Ohio Valley, swamping tobacco storage barns in Kentucky. The floodwater started showing up in his pipe tobacco. It tasted moldy. He could no longer smoke it. So he switched to cigarettes.

As any pipe smoker will tell you, breaking the pipe "chewing" habit is hard to do. That's one of the reasons people smoke pipes: They like to chew on the stem. So when Sveda switched he continued to chew on the stem, only this time there wasn't any; just the sticky cigarette paper. So Sveda found himself sucking on the paper.

One day in 1937 he set his soggy cigarette on his lab bench only to find that when he picked it up and puffed it tasted sweet. Why? Sveda wanted to find out. So he did the dance common to all pioneer sweetener scientists: He ran around and tasted every compound in his laboratory. Finally, he settled upon some crystals in a small dish. The chemical, sodium cyclohexylsulfamate, belonged to the cyclamate family of compounds. It was about thirty times as sweet as sugar but not nearly as sweet as saccharin, which was over five hundred times as sweet.

Drug and chemical companies wanted no part of a product that was less than one-tenth as sweet as saccharin. They turned it down despite the fact that cyclamates could be made sodium-free and had less of a bitter aftertaste than saccharin.

HOME SWEET HOME

Finally, in 1942, patent rights were assigned to Du Pont, where Sveda was employed. Du Pont assigned test and marketing rights to Abbott Laboratories. After further study, Abbott found the product better able to retain its flavor at higher temperatures than saccharin, and it tasted more like real sugar. Out of all of this came Sucaryl. To combat its bitter taste, saccharin was often combined with cyclamates. That sweet partnership ended in 1970 when cyclamates were banned from the United States when laboratory tests showed it caused cancer in rats. Both sweeteners can still be found in other countries. (Since people are choosy about their sweeteners, it's not very unusual to find people "importing" their favorite sweeteners from abroad in their airline luggage.)

ASPARTAME

NutraSweet, as aspartame has come to be known, was also discovered by accident. Chemists working in 1965 for the drug company Searle were trying to formulate a new anti-ulcer drug. In heating one batch of chemicals, Dr. James Schlatter inadvertently bumped some of the mixture onto the outside of the glass flask. Some of the powder spilled onto his fingers. Later, after licking his finger to pick up a sheet of paper, Schlatter noticed a very strong, sweet taste. At first he though he had spilled some sugar on his hand. But after remembering he had washed his hands just a short time ago, Schlatter figured the taste must have come from the chemical concoction he had formulated earlier. Knowing what was in the container—and therefore assuming it not to be toxic—he carefully tasted the powder and found the same sweet taste.

No mad dash around the lab was necessary.

18

Silly Putty: Science's Goofiest Discovery

First application—therapeutic use—Vaughn Ferguson's wife injured her hand. She was given Putty to exercise. Later came chair levelers, golf ball centers. Kids chewed it for gum. We would be routed out of bed in the middle of the night by hospitals trying to determine if it were toxic.

—GE scientists reminisce about the
discovery of Silly Putty, 1991

WHAT WOULD DISCOVERY be without controversy? A lot duller. Even if the discovery is of a scientifically useless product like Silly Putty. Silly Putty, that lovable taffylike toy sold in colored plastic eggs, was concocted by accident by chemists. But just who invented it and how it escaped the laboratory to toy store shelves is a matter of hot dispute in scientific circles.

The marketing arm of Silly Putty, Binney & Smith, tells its version of history this way. The year is 1943. World War II is raging throughout Europe. The war is eating up the best consumer goods. Newly invented nylon is gone from stockings and winds up in parachutes. The Japanese invasion of rubber-producing countries in the Far East has cut off the supply of natural rubber to the United States. Suddenly the war effort finds itself short of truck tires and boots.

Good old American ingenuity is asked to pitch in. The War Production Board asks American companies to try and develop a synthetic rubber.

Slaving away in his New Haven, Conn., laboratory, James Wright—known to his friends as Gilbert—is doing his best to aid the war effort. A Scottish engineer working for General Electric's silicone project, he is testing silicone for its rubbery properties. Silicone is a derivative of sand or silicon. That means it's an abundant material. If it could be "rubberized," the war machine would have an unlimited source of stretchy stuff.

The idea of using silicon to make rubber is not new. The idea dates back about eighty years to the decade between 1910 and 1920. British chemist F. S. Kipping, the pioneer of silicon-organic chemistry, was conducting "crazy" experiments: experiments that might or might not lead to anything—basic research—on the element silicon.

A SCIENCE FICTION DREAM

All common life forms on earth are based on the element carbon. Carbon can be found in the bodies of all plants and animals. In fact the word organic—used many times to mean life—is the chemist's nomenclature for carbon. Kipping wondered what would happen if silicon, a close cousin of carbon, could be substituted for carbon in organic molecules. What novel properties would those new substances have? (Many science fiction writers have wondered themselves; they often create alien worlds where life is based on silicon instead of carbon.)

These pioneering experiments produced unusual results. In some cases where silicon was substituted for carbon, the substances merely "gunked up" instead of forming free-flowing oils. In other cases they formed potentially useful products. Where carbon tetrachloride is flammable, silicon tetrachloride is not.

General Electric, a company very interested in rubber as insulation for wiring, hears of these pioneering experiments and becomes very interested. What if all the carbon in rubber could be replaced with silicon? Because carbon is the backbone of rubber, it burns easily. Because silicon is the main ingredient in common sand, it's inflammable. Could a silicone-based rubber product be produced, and if it is, would it be inflammable? The 1930s finds GE searching for the answer, but not without a twist of fate.

WORLD WAR II AND SILLY PUTTY

World War II and the shortage of natural rubber for the war adds impetus to the search for a synthetic rubber. Having years of experience with silicone, General Electric is chosen by the U.S. government, on the authority of Captain Hyman Rickover, to experiment on many silicone projects, one of them being silicone wire insulation for new electric submarine motors.

As senior staff member for the GE project, James Wright has been described by his colleagues as a romantic, a man willing to toy with new ideas. Trying to find the perfect kind of silicone for practical use, Wright tinkers with various substances. He learns how to build the silicone rubber "skeleton," the backbone of the silicone molecule. His son-in-law, James Marsden, works alongside him to improve the skeleton and by accident discovers a way of making the gooey silicone a harder, more useful rubber.

Continuing his "crazy" experiments Wright accidentally tosses a little boric acid into silicone oil in a test tube. Expecting the mixture to produce hard rubber, Wright is astounded. It's gooey. Taking some of the stuff out of the test tube, he throws it onto the floor and to his amazement, the gob bounces back. (Scientists will later say that the silicone has become "polymerized.") "Bouncing putty" is born.

WHAT GOOD IS BOUNCING PUTTY?

Elated, Wright discovers that bouncing putty can be stretched and pulled. The laboratory is in an uproar. One of the chemists stands in the hallway with a handful of the new substance, rolling it, dropping it on the floor, and shouting, "Look at it bounce." But what can be made with it?

"Well," says the chemist, "you can use it to roll, drop on the floor, and say, 'Golly, look at it bounce!' "

Looking for some useful suggestions, General Electric, in 1945, sends samples to engineers worldwide. Unfortunately it becomes evident that one thing bouncing putty is not good for is making truck tires or galoshes. No practical use can be found. So much for the war effort.

But bouncing putty is a lot of fun. Not only will it bounce, but the strange-looking pink gorp can be stretched and squeezed to almost any shape. By 1949 bouncing putty has circled the globe to become the subject of cocktail party small talk. Scientists and technologists can't stop talking about it.

Attending one of these affairs is Ruth Fallgatter, owner of the Block Shop toy store in New Haven, Conn. She hires marketing consultant Peter Hodgson to produce her catalog. The two decide to run a verbal description of bouncing putty in the catalog on the same page with a spaghetti-making machine and a toy steam roller.

Sight unseen the bouncing putty, costing two dollars and packed in a clear compact case, outsells every item in the catalog save one—Crayola crayons. Despite the success, Fallgatter declines further interest in marketing it.

SILLY PUTTY SUCCESS

Hodgson steps in. He borrows $147 for a batch of the gooey stuff and packs one-ounce wads of it in clear plastic eggs selling for a dollar. It's introduced at the 1950 International Toy Fair in New York as Silly Putty and is greeted with a large yawn from the toy experts.

Undaunted, Hodgson persuades Neiman-Marcus and Doubleday bookshops to carry the plastic eggs (now shipped in surplus egg boxes supplied by the Connecticut Cooperative Poultry Association).

Then comes the big break. In August a writer for *The New*

Yorker drops into Doubleday in New York and is mesmerized by Silly Putty. Silly Putty is featured in the "Talk of the Town" section of the magazine. The results are astounding. Hodgson receives orders for more than three-quarter million eggs in three days.

In 1968 Silly Putty goes to the moon with *Apollo 8* astronauts.

BREAST IMPLANTS

Results from early wartime experiments with silicone insulation indicated that silicone would be a good insulator. But further work showed it to be a flop. Silicone rubber adhered poorly to wires; they became brittle and broke. The electrical motors simply did not work. Teflon, another wartime product, would deliver on all of silicone's promises (see Chapter 14, "Teflon: The Top-Secret Discovery").

However, a silicone gasket for battleship searchlights and silicone film for keeping electrical circuits dry worked quite well. And along the way, the GE silicone project led to silicone gel used (controversially) in plastic surgery to enlarge breasts.

AND NOW FOR THE CONTROVERSY

So much for the manufacturer's side of the story. Was James Wright the first to invent the stuff as Binney & Smith would have us believe? Not if you ask Earl Warrick. Warrick claims that he and R. R. McGregor invented the material earlier while working at Mellon Institute. And he has the facts to back the claim up.

A patent for bouncing putty was issued to Warrick and McGregor years before Wright received his. McGregor and Warrick received theirs on December 2, 1947, and it was assigned to Dow Corning Corporation. Wright received his on February 13, 1951, and it was assigned to General Electric.

Cartoon from *Collier's,* December 1944, showing GE scientists experimenting with bouncing putty.

Warrick says the magical material *was* invented by accident, but by an event that occurred in *his* lab. "We were trying to do something else," he says, "make silicon rubber." Working for Dow Corning, Warrick put a mixture of boric acid (boron trioxide) in a test tube with an oily material called Corning 200 fluid, which has about the same viscosity as mineral oil.

"We put it in, heated it in an oven overnight, and lo and behold the boron oxide linked to polymers." That's chemist talk describing the creation of rubber.

"We got a rubbery looking material, we bounced it around the lab and amused ourselves and our visitors by pulling it apart. But when we put it in water the boron came out."

All that was left was the original oily fluid. What good was a rubber that came apart in water? As a toy "to astound visitors by bouncing it off the ceilings and walls of our laboratories."[1]

"We kept it around as a laboratory curiosity. The company made some and gave it to salesmen to show around."

HODGSON MEETS DOW CORNING

According to Warrick, Dow Corning showed it to Peter Hodgson. He took the stuff, put it in little eggs, and called it Silly Putty. "One smart thing," says Warrick, "was he copyrighted the name. A copyright goes on forever. Our patent was only good for seventeen years." By the time Dow Corning could mount a marketing campaign to sell the toy, if it wanted to, the patent would have run out. And because Silly Putty could be made so easily, and with so many similar ingredients, there was no compelling reason for the chemical company to claim priority. Another company could change the formula slightly and create its own.

That's exactly what happened, according to Warrick. James Wright created the same product but via a different stroke of luck. He was mixing into the Corning 200 fluid various fillers and heating them. One of the fillers was boron nitride, a heat-resistant material, which GE hoped would create a greaselike, heat-resistant rubber. But when the mixture was heated, out popped bouncing putty because boron nitride has an impurity in it called boron oxide, which is similar to what Warrick was adding himself.

"They did it five years after we did," says Warrick, "but because they did it by a different route they got their patent. But it was the impurity in the boron nitride that did it."

As for Hodgson, chances are he bought his supply of Silly Putty from everybody: Dow Corning, General Electric, or anyone with a simple chemistry lab.

Warrick keeps a can of the original putty, nestled in its original canister, at his home in Michigan. If General Electric has any of its original bouncing putty left, it can't be found.

Now, doesn't that set the record straight?

19

....

The Computer That Saved D-Day

Never have so many owed so much to so few.
—WINSTON CHURCHILL

HOW DO YOU hide a building full of electronic computers each made of twenty-four hundred vacuum tubes and eight hundred electronic relays? Obviously, very carefully, since their existence was kept secret for over thirty years.

Spring 1944. The German army is anxiously awaiting an invasion of occupied Europe they know is coming. But where? The Allies work on keeping a tight lid on the invasion site: Normandy. Fearful that the enemy will discover the secret plans, the Allies try to divert Nazi attention from Normandy, even planning and carrying out clever ruses. Breaking the code on a secret telegraph link between the German Commander-in-Chief West, General Karl von Runstedt, and Berlin, the Allies learn that the Germans are not about to be fooled and are feverishly building up their defenses on the Cherbourg peninsula. This top-secret information is just what the Allies need to know. They quickly change the details of their invasion plans. The rest, as they say, is history.

The heroes of the invasion that turned the tide of the war are not only the soldiers, sailors, and fliers who land on the beaches but super-secret electronic computers whose existence remained secret for thirty years following the war. Unbeknownst to the rest of the world, back in England the world's first vacuum-tube computers, named Colossus, succeeded, working day and night, at deciphering the top-secret German military code. And by the end of the war, these computers contributed not only to the military

ENIAC. (Electronic Numerical Integrator and Calculator) was a collection of twenty electronic adding machines that operated jointly as a computer. ENIAC was designed to calculate ballistic equations for weapons firings.

triumph of D-Day but to every major battle in European theaters: North Africa, southern France, Italy, and eastern Europe.

In contrast to their widespread use in business today, the grand-daddies of modern-day computers were built in the 1940s not for routine bookkeeping or word processing but to help win the war by breaking secret German codes. And while the Eniac computer built by the University of Pennsylvania has long been regarded as the world's first, Eniac may actually have an older brother: the British Colossus. Writing in *IEEE Spectrum,* Glenn Zorpette pointed out that some historians now argue that Eniac was not the world's first computer—or even its second—but the eleventh if all ten Colossus machines are counted. Yet hardly any historians have ever credited the British with inventing the first computer, and for a good reason. Colossus has received such little notice due to the British penchant for secrecy. Due to the ultra-secret nature of their business, no cryptographer had admitted Colossus existed until word leaked out in 1976. But in the intervening years, an incredible picture is emerging of just how important computers, including Colossus, were in winning the war.

BREAKING THE GERMAN CODE

Almost from the beginning, British cryptographers enjoyed great success in breaking Nazi codes. The mainstay of German intelligence was a machine called Enigma. Looking a lot like a small, old cash register with hand crank and dials, Enigma was the coding and decoding machine of all German troops and espionage agents. And it was an ancient machine. Invented in 1919 by two Dutchmen, it was produced in Berlin and sold to the public. Various models were even patented in the United States in 1928, 1929, and 1933. The Germans so liked the machine that they modified it to their own secret military standards in the 1930s. But by then its basic operation had become so widely known that three Polish mathematicians, with the help of military manuals smuggled out by a spy, were able to break the code of the military version in 1933.

Building a replica they called "bomby" (from the name of an ice cream treat, *bomba,* the mathematicians were said to have been eating at the time), the Poles secretly stole German secrets with the machine until 1939. By the time of the German blitzkrieg of Poland, British and French agents had copies of the Enigma sitting on their desks.

The British improved the machine significantly, renamed it the "bombe," and showed it off to American cryptographers. The Americans made their own—some say even better—copies of the bombe. And together the Allies began an incomparable intelligence coup. By the end of 1943 and through the rest of the war, British intelligence was decoding ninety thousand messages a month sent by German Enigma machines.

SINKING THE <u>BISMARCK</u>

Although the public never knew it until decades after the war, cracking the Enigma code let the Allies win major battles of World War II. When the "unsinkable" German battleship *Bismarck* was finally hit on May 26, 1941, by British aircraft and sought to limp

Colossus. Before ENIAC, the Colossus was built (1943) by the British government to decode secret German war communications. Here Colossus is shown being operated by members of the Women's Royal Naval Service.

home for repairs, a frantic search to find her proved fruitless. His Royal Majesty's greatest ships and planes could not find the ailing and vulnerable ship. But British cryptanalysts did. Having already read the secret transmissions of the German air force for months, they decoded German messages to learn the *Bismarck* was heading for Brest for repairs. Searching the Brest coastline, British aircraft spotted her eight hundred miles from the safety of the French harbor. At 9 A.M. the next day, British battleships *Rodney* and *King George V* opened fire on the German battleship. An hour and a half later, the star of the German fleet went to the bottom with two thousand of her crew.

This was only the beginning. By the middle of the war, in the summer of 1943, seventy German U-boats went to watery graves, their positions given away by intercepted radio signals to home. Allies were regularly sinking German supply ships headed for General Erwin Rommel's Afrika Corps in North Africa. They even knew of Rommel's ill health via intercepted radio messages from his doctor.

In the Pacific, American cryptanalysts were having their own field day. Having already broken the Japanese codes before the war, American intelligence was able to routinely intercept and decipher just about every Japanese message they could pick up. This intelligence coup was instrumental in secretly finding the Japanese fleet and winning the Battle of Midway in 1942, which broke the back of the Japanese navy and turned the tide of the war in the Pacific.

Breaking the Japanese code paid dividends regarding Berlin, too. As the Japanese ambassador to Germany was apprised of German intentions, Americans eagerly tuned in to the latest German thinking as German intentions were radioed to Tokyo by the Japanese embassy in Berlin.

The Allies were overjoyed with the wealth of intelligence they were reaping. Just about every word sent by wireless could be deciphered. It was an intelligence bonanza. But it wasn't enough. The British had their hearts set on going in even deeper. Decoded radio communications were helpful, no doubt about that. But what

British cryptanalysts thirsted for was the Holy Grail of cryptography: some way of deciphering the most secret of all messages, those that traveled by Teletype between Berlin and the top German army commanders scattered throughout Europe—messages sent by Hitler himself.

This was the ultimate challenge, and success required a level of sophistication never achieved before, a method that would change the face of cryptography forever. Success required inventing a machine that would dwarf Enigma and all those that had come before. Waiting to be born was the machine of the future: the high-speed electronic digital computer.

German Teletype code was produced by a machine built by the German company K. Lorenz in Berlin. The Lorenz machine produced a code dubbed Tunny by the British (British for "tuna"). The British threw every resource they could into breaking Tunny: scores of cryptographers and engineers, hundreds of assistants working double shifts for months on end.

Encoding a message was relatively simple. The Lorenz machine would employ a widely used encoding method known to all cryptographers. Here's how it worked. The message was broken up into a string of letters. Each letter of the code would be paired with a second letter to produce a third letter. The list of second letters was called the "key stream." So, for example, the word "boy" would be broken down into its three letters. The "b" would be matched with a random key stream letter—let's say "h"—to produce, let's say, "j"; the "o" would be matched with another random letter—let's say "g"—to produce "b" and the "y" with a third random letter, let's say "a," to produce "t." The resulting coded word, "jbt," would be transmitted to agents or military personnel in the field. The coded word was just one of many strung together and producing a sequence of letters looking like gibberish. If you, as an authorized receiver, had the secret code book containing the key stream, you would reverse the process and decode the word. You would take out your key stream and because you had the secret code book you'd know how to line up the key stream side by side next to the code to decode each letter.

The key to cracking the code was in re-creating the key stream. The key stream, to be secure, had to appear to a code breaker to be absolutely random. No detectable patterns to it. No beginning, no end. And the Lorenz machine was able to crank out a code whose key stream appeared to be totally random. Lorenz produced a key stream with no recognizable pattern in the course of making over 16 *quintillion* numbers—that's 16,000,000,000,000,000,000! No enemy cryptographer could hope to crack that code working with a pencil and paper.

THE SEVEN PERCENT SOLUTION

How could anyone ever hope to break that code? Enter one mathematical genius with a bright idea. William Friedman, in 1920, was employed by Chicago millionaire George Fabyan. At the age of twenty-nine, the young cryptographer had observed that if you take two long strings of English-language text and line them up side by side to form pairs, almost 7 percent of the pairs would be matches—two Ts or two Es, for example. But if you paired up randomly selected letters, matching pairs would occur just 4 percent of the time. The difference in percentages is the result of the fact that some letters in the English language occur more often than others, especially E and T, the two most common.

To the untrained eye, a 3 percent difference is hardly worth noticing. But to a cryptographer, it's nirvana. If two long strings of encoded text could be somehow lined up together so that the key stream for each text was aligned too, the matches would jump to 7 percent. It didn't matter if the text were encoded; the letters would appear as coded matches.

That 3 percent difference is enough to make these letters stand out. Once chosen by a cryptographer, these known letters form the backbone of a good guess, like the cryptograph puzzle in the daily newspaper. Given enough time, and guesses, the puzzle—and the code—can be broken.

So much for theory. The real crunch comes in trying to line up the key streams. With a key stream measuring in the quadrillions, there's not enough time on earth for all the cryptographers and their pencils and paper to make it work! It became clear that humans alone could not solve the problem. What was needed was a machine. One that could repeat the comparison over and over again until it found the key. And one that could do it thousands of times a second. There was just one slight problem. Such a machine hadn't been invented yet.

A LUCKY BREAK

The job of building such a machine fell on the shoulders of Maxwell Newman, a mathematician at Cambridge. Newman was to work out of the Government Code and Cipher School at Bletchley Park, housed in a converted mansion northwest of London.

And the British might never have invented such a machine without an unbelievably fortunate mistake made by an unknown German code machine operator in the winter of 1941–42. As Zorpette pointed out in *IEEE Spectrum*, the British have never divulged the exact nature of the mistake, but it allowed the British to determine what the insides of the coding machine looked like (twelve encoding wheels, each containing a differing number of pins) and how each of the wheels turned one another to encode a message.

To the Bletchley crew, it was as good as having the machine themselves. With knowledge of the inner workings of the machine, breaking a code would amount to deciphering the patterns of each of the twelve wheels, and then deciphering the initial settings of the wheels when the code was formed. This was the mechanical equivalent of lining up two coded texts side by side and visually finding a match. It was no mean feat. But it at least allowed itself to be solved by machine, if only one could be found.

A CARTOONIST'S DREAM

Newman's first attempt at creating an automated code cracker revolved around two paper Teletype tapes. One tape's holes contained information about the twelve wheels and the way they were configured. The other tape contained the message to be decoded. Both tapes were fed into a single Teletype machine.

Put into operation in September 1943, the machine was called the Heath Robinson. Named after the cartoonist W. Heath Robinson, the British version of America's Rube Goldberg, the decoder lived up to its name as a noisy and unreliable contraption. In their haste to design the fastest operating machine possible, the engineers had sacrificed reliability for speed. While the tape reader could read two thousand characters per second, it had the nasty habit of chewing up the very paper it was reading. Since each paper tape had to read through the mechanism thousands of times, the tape didn't last very long.

Even so, the machine enjoyed a modicum of success, decoding a few German Teletype messages, enough to convince the government that they had the kernel of a good idea and to proceed further. If only some way could be found to eliminate the unreliable and destructive tape reader. What was needed was a way of loading in the tapes just once and reusing the data—the equivalent of modern-day computer memory. It didn't exist. But it was about to be invented.

In northwest London, at the Post Office's Dollis Hill Research Station, Thomas H. Flowers had been working on a project called Colossus, an attempt at modernizing the phone system by replacing some of the mechanical switches with electronic ones. In 1939 he had plans to actually build the first fully electronic telephone exchange powered by vacuum tubes. World War II brought that dream to an end. Fortunately, Flowers found himself working alongside the Heath Robinson crew and rubbing elbows with them. He suggested that perhaps work he had done on the phone system might be applied to the problems of Heath Robinson.

Specifically, Flowers suggested replacing one of Heath Robinson's tapes with the Colossus internal electronic memory system. This would eliminate the problem of reading one of the tapes at high speed and replace the finicky tape reader with the reliability of Colossus. Like all great innovative ideas, this one was not overwhelmingly supported. Flowers's colleagues pointed out that vacuum tubes were not very reliable. Over long periods of time they burned out. And the Colossus Mark I machine, which Flowers had proposed, would be composed of fifteen hundred tubes. Imagine the maintenance problems in finding and replacing burned-out tubes in such a machine.

NEVER TURN IT OFF

That's when Flowers suggested one of those simple ideas that change history. His response: Never turn the machine off. Vacuum tubes wear out because we turn them on and off too much. The spike of electrical current surging through the tubes during power on and power off is more damaging than simply letting them burn twenty-four hours a day.* So maintenance on thousands of vacuum tubes can be reduced to a manageable level by simply letting them run continuously. And that's what was done.

The Colossus Mark I went on line that year, in December 1943. It ran twenty-four hours a day. In June 1944 an improved computer, the Colossus Mark II, came into service containing twenty-four hundred tubes augmented by eight hundred electromagnetic relays. It too was never turned off. Both were highly reliable.

Operating in parallel, a number of Mark IIs could read twenty-five thousand characters per second. The electronic brains of the Mark IIs could spit out sequences of numbers like the cogs of the wheels of the encoding machines. These key streams could be adjusted by an operator.

Running a code through Colossus led to a 90 percent success rate in deciphering Tunny messages. But even such a machine, which took up a whole room to itself, needed help to finish the job. Cryptanalysts took over and worked the last stages of decoding by hand. The whole procedure took a few days to decode a mes-

---

* This idea has carried over into the modern solid-state world of transistors. The only time industrial electronic equipment is turned off is for maintenance.

sage, even when the wheel settings were known and Colossus was doing its best.

INVALUABLE AT D-DAY

Deciphered Tunny messages provided valuable intelligence in just about every theater of war in Europe. But Colossus's most shining hour came in helping plan and execute the invasion of Normandy. Breaking the Tunny code of the German Teletype link between Berlin and commanders in the field, Colossus told the Allies how

Brattain's Lab Notebook. On Christmas Eve 1947 Walter Brattain recorded in his notebook the historic event of that month: the operation of the world's first transistor. Brattain and a fellow scientist at Bell Labs, John Bardeen, succeeded in producing the first solid-state amplifier, built on a silicon slab. The era of the vacuum tube was about to end, and with it came the dawn of the age of Silicon Valley.

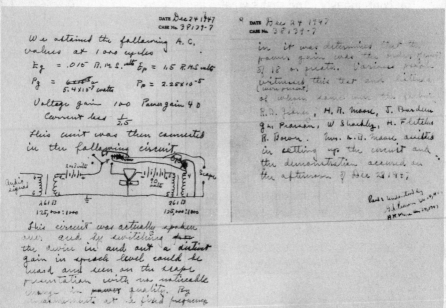

German officers were planning to counter an invasion from early March 1944 until four days after D-Day, June 6. While British intelligence still remains secret about what was learned, they do admit to learning that the Germans were heavily fortifying the Cherbourg Peninsula. Allied invasion plans were effectively changed due to this intelligence coup.*

Colossus was such a secret project that even the people who designed and built the electronics didn't know why or what it was for. Historians now believe that ten Colossus machines were built during the war. Ignorant of the existence of Colossus, historians had bestowed the title of the world's first computer on the University of Pennsylvania's ENIAC.

British computer experts now like to think of ENIAC not as the first computer, but the eleventh.

THE ENTOMOLOGY OF COMPUTER BUGS

Where a calculator on the ENIAC is equipped with 18,000 vacuum tubes and weighs 30 tons, computers of the future may have only 1,000 vacuum tubes and perhaps weigh 1¹/₂ tons.

—*Popular Mechanics*, March 1949

Computer folklore says that when ENIAC, the country's first general-purpose electronic computer, was finished in November 1945 it had so many warmly glowing vacuum tubes that moths were a constant nuisance. They commonly flew into the machine,

* The details of the Normandy events remain secret even to this day. What little that's been made public is described by journalist Glenn Zorpette in an article ("Breaking the Enemy's Code") published in the September 1987 issue of the *IEEE Spectrum* (IEEE, New York, pp. 47–51). Zorpette details how the lid on the secret was opened just a bit in March of 1986 in a London speech by Sir Harry Hinsley, a former British intelligence officer who was intimately involved in breaking the code. All Sir Harry would say was that "the outstanding contribution" made by breaking the code was that intelligence arrived in time for significant, effective changes in the invasion plans. What those changes were, he would not say.

according to the stories, and a single moth in a futile mating dance with a vacuum tube could bring ENIAC to a sudden halt.

Is there any wonder, then, how the word bug came to describe a problem that upsets a computer and debugging to describe the process of fixing those problems?

It's a great story. And one that is only slightly ruined by a few facts. First, ENIAC was housed in a sealed, air-conditioned room. A moth didn't stand a chance of getting in. Second, while ENIAC did malfunction at times, technicians can't remember ever using the word bug. The word intermittent was a favorite. (Intermittent was, and is, commonly used to describe electrical wiring that is loose. It's on, it's off, it's on again.)

The First Computer "Bug"? This moth, taped to a technician's log book, shut down a room-sized computer when it flew into an ancient machine at Harvard University in 1945.

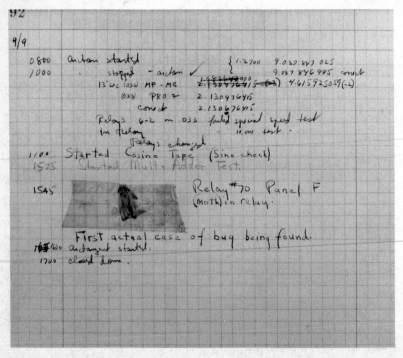

THE FIRST ACTUAL BUG

So much for the bad news. The good news is that a moth *was* involved in crashing a computer on September 9, 1945, just two months before ENIAC began operating. This time the killer moth flew into the works of the U.S. Navy's Mark II, a room-sized calculator chockful of switches and relays instead of vacuum tubes. In an un-air-conditioned room at Harvard University, the moth made its way into an open window and thence into a relay. Jammed between the contacts, the moth was beaten to death by incessant toggling of the relay switch.

Only when the moth brought the enormous calculator to a halt did the operators look for the problem. Finding its broken body, they removed the moth with a pair of tweezers, entered the incident in a log book, and Scotch-taped the moth to the entry to silence the snickers of all the Doubting Thomases in the lab. (If you're one of them, see a picture of the actual log entry—complete with moth—on page 180.) After that, whenever the Mark II was down for repairs, programmers described the process as debugging.

WHEN COMPUTER VIRUSES ARE GOOD FOR US

Open the pod door, Hal.
           —2001: A Space Odyssey

Many years ago when I started tinkering with computers, the moniker hacker was worn with pride. In those days computers were very ugly, crude, and unforgiving machines. They defied anyone to enter their souls and emerge victorious. Hackers were programmers who could overcome the layers of numeric gibberish and actually make computers perform useful work for the uninitiated masses. When a computer refused to run your program, hackers could by sheer brain power find a way around it. They were the Davy Crocketts of a frontier at a time when the term user friendly had yet to be discovered. I prided myself on being a junior hacker from the mere fact that I could program a computer the size of a building. Unfortunately, hacker has recently become synonymous with thief, as in breaking and entering. Sensational front-page news of their unauthorized use of computer networks has tarnished the image of computer mavens. Little notice is paid to the hackers wearing white hats who tirelessly worked to capture and prosecute their wayward compatriots.

The same can be said for computer viruses. In 1989 a computer virus swept through the Department of Defense's Arpanet system, one of the country's biggest and most used computer networks. Like an unchecked epidemic the virus infected one computer after another, contaminating each system on contact.

In reality it created no more physical damage than a bad cold. The patient, though inconvenienced by time lost from work, and poorer from out-of-pocket "physician's costs," recovered completely. And as in most cases of mild diseases, the victim emerged a healthy being, equipped with a stronger protective shell, ready to stand guard against the next onslaught, if and when it comes.

Contrary to the banner headlines of the day, the virus scandal perpetrated by young computer programmer Robert T. Morris was not really as earthshaking as one would imagine. It involved no real breakthroughs in technology, no real threat to secure data, and the virus was actually blocked from entering some computer systems that were prepared to prevent just such a "disease" spread.

In fact, what you probably haven't heard much about is the other side of the story: how this virus is really the descendant of a new type of computer software designed to enhance networks instead of harming them.

It's a well-known secret among programmers that networks, including Arpanet, have been purposely infested with viruses for almost two decades. But programmers have other names for them. They call them worms. Worms are constructive, useful little creatures. It's only when they are harnessed for mischief that these programs are called viruses.

THE BLOB

Worms trace their heritage back to two researchers at the Xerox Palo Alto Research Center (PARC) who proposed a brilliant and novel computer programming idea based on the science fiction movie *The Blob*. Later, writing in the journal *Communications of the ACM*, John F. Shoch and Jon A. Hupp described how the program "started out running in one machine, but as its appetite for computing cycles grew, it could reach out, find unused machines, and grow to encompass those resources." The worm, as this program was dubbed, would become active at night when the computers on the network were most idle. (Shoch says that one researcher described these nightly visitors as "vampire programs.")[1]

In the middle of the night, such a program could mobilize hundreds of machines in one building; in the morning, as users reclaimed their machines, the "blob" would have to retreat in an orderly manner, gathering up the intermediate results of its computation. Holed up in one or two machines during the day, the program could emerge again later as resources became available, again expanding the computation.

The beauty of these nocturnal computational doings is the ability of the worm to slither among computers and harness many machines to work on a single problem too big to be handled by just one computer.

The concept is elegant. The segments of the worm are in constant communication with one another. If one segment fails, the other pieces find a free machine and add it to the worm. As machines join and then leave the project, the worm appears to move through the network. When it finds an idle machine, it replicates in a controlled and useful manner.

The worm does not do the actual "thinking." It's there to gather and maintain the segments of the worm. The definitive programs that do the work are built into the worm, coming along for the ride like the warriors atop the worms in the novel *Dune*.

THE CREEPER AND THE REAPER

In the early days of worm design in the 1970s one of the first programs to move by itself through the Arpanet was the Creeper. Designed as a purely demonstrational program of worm power, the worm would start to print a file, stop, transfer to another machine, and start printing again. Success with Creeper led to Reaper, which not only moved through the network but replicated itself while looking for copies of Creeper, which it terminated.

McRoss was a simulated air traffic control program run between computers. Each computer ran one part of the simulated air space. As planes flew in and out of new air space, they were handed off from computer to computer. In later models, even the air space being simulated could be handed off. This meant that during the move, the machines had to suspend simulation, reestablish communications with other parts of the simulation, and pick up where they left off. With no loss of functionality.

The flexibility of these demonstration programs looked like the answer to a computational dream that finally bore fruit.

For many years, mathematicians had been trying to solve one of the greatest problems in their field: how to factor a hundred-digit number, that is, how to find the smaller numbers that when multiplied together, produce the behemoth hundred-digit monster. Besides endowing bragging rights on the lucky person who solves this problem, the solution would have immediate benefits in the world of encryption: secret coding of computer data or voice communication.

Just recently the infamous hundred-digit problem was solved using a worm program. Because no single computer was powerful enough to solve the hundred-digit problem itself, the program was split into pieces and segments of this total worm were distributed among many machines interconnected in a network. By turning the worm loose in a building full of networked machines, the worm was able to creep among the hardware, drawing whatever computational or hardware support it needed from each machine.

THE KNOWBOT

Another innovative use of the worm concept is called the knowbot. Knowbots are little programs that go around the networks performing useful and time-consuming (for humans) deeds. And while viruses, cousins to knowbots, are unwelcome visitors, knowbots represent the kinds of worms that are well-known entities in the system that you expect to show up and try to do useful things on your behalf.

You could tell a knowbot to perform as simple a task as "Find me two novels by Michener I haven't read" or as time-consuming a job as "Find me a book (in any library) about decision-making strategies of international high-tech corporations." The knowbot could go through a network of libraries around the world searching for this hard-to-find information. It would knock on the doors of computer libraries in foreign countries and be welcome. The machines would be prepared to accept it.

Knowbots are still just a concept that universities around the country, the storehouses of hard-to-find information, might soon adopt. The goal is to create a digital library system—documentation and data that could be read by computers—and to do for the world of multiple data bases what was done for the field of networking back in the seventies.

Worms like these are designed to be among the most useful and creative pieces of computer programming to come along. However, this global view of networking—and even the local office variety—may have trouble surviving if people don't understand the philosophy behind networking and are unnecessarily panicked by this latest computer virus. This was not the AIDS epidemic of the computing world.

Computer networks are like communities where people live and work. Part of the social convention in any community is that we don't break into other people's homes and rummage around, even if we don't take anything. The same is true of the computing community. It's an unwritten (if not legal) custom that someone else's data, be it stored in a mini, mainframe, or microcomputer, is off-limits to unauthorized users.

Hacker Morris didn't find an evil new way of breaking and entering. He merely violated a social trust. Because networks many times are like houses that keep their front doors locked but their back doors open, Morris, in effect, merely entered through the back door. He exploited loopholes in the system (called trap doors) well-known to many programmers and tread where they would not go.

While he needed to break a few computer passwords to enter, what he breached, more importantly, was the social contract of the community.

This is not solely a computing problem. Credit card companies must trust their card holders not to abuse the system with phony cards. So placing the entire fault on network security is in one sense a red herring. It would be very easy to make networks much more secure by loading on layer after layer of computer passwords and IDs.

The concept behind a network is to keep it friendly enough that people will like to learn and use the system. Too much security would defeat this purpose. But given human nature, a certain amount of security will always be desirable.

While the tug of war between security and user-friendliness goes on, there is at least comfort in knowing that the viruses in question have not cropped up suddenly like a silicon Andromeda strain. They can trace their lineage to a well-known and highly respected family that at times is known to spawn, as we all do, certain black sheep members.

20

....

The Submarine: Made in New Jersey

*Without celebration the <u>Holland</u>, the little cigar-shaped vessel
owned by her inventor, which may or may not play an
important part in the building of the navies of the world
during years to come, was launched from Lewis Nixon's
shipyard this morning . . .*

—*New York Times*, May 17, 1897

THE ERA OF the submarine was born one spring morning in New
Jersey almost a hundred years ago. John Holland, the sub's inventor, knew he was making history that morning, but the kind of
history he had in mind was not exactly what occurred. Holland had
dreamed about underwater boats his whole life. Born in Ireland in
1843, Holland sketched designs for submarines at the age of seventeen. But it wasn't until he emigrated to the United States in the
1870s and settled in Paterson, N.J., as a school teacher that his
ideas took shape.

Paterson, full of machine and boiler shops for tinkerers, was a
town that had a reputation for attracting inventors. Samuel Colt,
inventor of the Colt revolver, and Thomas Rogers, steam engine
builder, flocked there. Although it was inland, the nearest navigable
sea water being twelve miles away, Holland chose the site as a
place to get serious about building underwater boats.

THE NAVY SAYS NO

Holland's job as a school teacher meant he had to design his subs in
his spare time. In 1875 he offered the United States Navy the plans

for a submarine that could be automatically maneuvered under water. He waited for a reply that never came. The navy ignored him. Nobody was interested in building submarines with plans submitted by the energetic young man with the thick glasses and walrus mustache. Perhaps they didn't believe that any school teacher working in his spare time could come up with anything of value.*

But the feisty Irishman would not take no for an answer. Borrowing money, he built his first submarine in neighboring New York in 1878. It was a tiny one-man experimental vessel, fourteen feet long, named *Holland I*. Hauling the sub back to New Jersey, he launched her into the Passaic River. She dropped like a rock to the bottom. Holland saved his life by swimming to the surface. He had erroneously based his buoyancy calculations on salt water, not fresh.

YOU CAN'T KEEP A GOOD MAN DOWN

This mistake earned the vessel the moniker Holland's Folly. Undeterred, Holland raised her from the bottom and took another dive a month later. This trip was a tremendous success. Holland piloted the vessel under water for an hour at a depth of twelve feet, achieving a speed of three and a half miles per hour.

Eager to move on with a better design, the inventor stripped the *Holland I* of her turret and other important parts and left the hull under water. The hull of the sub would be difficult to conceal in town. Better to let her secrets rest at the bottom of the Passaic River where rival inventors couldn't snoop.†

Holland was elated. Now on to bigger and better things: a vastly improved model. Broke and with no profit to show for his work,

---

* Thirty years later, another unknown genius working on a project in his spare time while holding down a menial job in the patent office would offer the world the theory of relativity. His name was Albert Einstein.

† The *Holland I* was located using a magnet under water in 1927 and brought to the surface. She and her sister ship, the *Fenian Ram*, are on exhibit at the Paterson Museum.

Holland once again turned to the navy, and once again the navy turned him down.

He turned to his home country for help. The Irish Republican Brotherhood, an anti-British group known as the Fenians, came to his aid with a check for $23,000. The Fenians hoped to use "their" sub to sink British shipping. The result was the *Holland II*, a new vessel twice the length of the first and dubbed the *Fenian Ram*. Launched in 1881, the *Holland II* looked like a metal cigar. Weighing nineteen tons and equipped with a one-cylinder internal combustion engine, her hull measured thirty-one feet long, six feet wide, and seven feet deep.

THE BIRTH OF THE MODERN SUB

But you can't tell a sub by its cover, and the *Holland II* was no exception. On her maiden voyage, she submerged sixty feet off Staten Island and stayed at the bottom for an hour. Rising dramatically to the surface, Holland knew he had a brilliant success. He had designed into his sub all the principles of diving, surfacing, and navigating that would remain unchanged in modern submarines. The periscope was all that remained to be invented.

Holland was so proud of his achievement that he could be found showing off his sub all over New York harbor. No one knew where Holland would be popping to the surface next. It scared a lot of other captains. It eventually cost him his submarine. With no periscope to see the way, the *Holland II* collided with a Weehawken ferry boat and sank in June 1882.

Luck still rode with the captain; Holland and the ferry passengers escaped unhurt, and Holland raised the sub a few days later.

Now Holland had a successful boat and nothing to do with it. He continued to improve her design. In May 1897 the *Holland VI* slid into the waters of the Arthur Kill, between Staten Island and New Jersey. Weighing seventy-five tons, the little boat was just fifty-three feet long, but she advanced the state of the art. Equipped with a gasoline engine for surface running and batteries for under-

Sub Commander. John P. Holland standing in the conning tower of his submarine, *Holland VI*, April 1898, Perth Amboy, N.J.

water power, she was the first battery-powered submarine. Almost a third of her weight—twenty-two and a half tons—was taken up by storage batteries. Under water, her electric motor turned a clutch-driven propeller to a speed of eight knots. On the surface, a small gasoline-powered engine took over and pushed her along at six knots.

The *Holland VI* made the public take notice. The *New York Times* was on hand to document her first launching. A technical publication of the day, *The Electrical Engineer*, described her virtues: "It's only a moment's work to shut down the gas engine and with electric power dive to a safe depth and run forty or fifty miles . . . there is plenty of electric light, and plenty of air stored in the compressed air tanks."[1] She contained a bilge pump, an air compressor, and two electric motors to run them.

THE NAVY BUYS A SUBMARINE

What more could the navy want? Even the U.S. government couldn't ignore Holland now. On October 2, 1900, the United States Navy bought the *Holland VI* and christened her the *USS Holland*. John Holland never built another submarine. But he believed that his invention was so revolutionary that the submarine would "end naval warfare." He foresaw how the sub could be turned into an offensive weapon. One of his notebooks contains data of an experiment he did under water: "U.S. Army rifle when fired under water projects a bullet four feet one and eleven-sixteenths inches—a range of 3,182 feet in air; one-quarter inch in water is equivalent to 16 feet in air." In the twenty years following Holland's Folly, John Holland probably spent more time submerged than any other person in history. He never made much money from his success—his salary peaked at ninety dollars per week.

Despite this success, Holland's biggest disappointment was his inability to bring to fruition his other lifelong dream: a flying machine. Looking strikingly like a modern-day helicopter, the design was rejected by the U.S. Patent Office in 1904, citing "the inoperativeness of the device for any useful purpose."

21

....

The Dream That Won the Battle
of Britain

*And Pharaoh said to Joseph, "I have had a dream, and there
is no one who can interpret it; and I have heard it said of you
that when you hear a dream you can interpret it."*

—GENESIS 41:15

CENTURIES BEFORE CARL Jung built his reputation on dreams,
biblical scholars pointed to Joseph, the amazing dream machine. In
both cases, dreams were believed to be the playground for the
unconscious mind, where problems unsolvable during waking hours
can magically find solutions in the unconscious mind.

Nowhere in the annals of the Bible or biophysics is there a story
more fascinating than the dream that helped change the tide of
battle in World War II. The hero is a twenty-nine-year-old engineer, David B. Parkinson, employed by Bell Telephone Laboratories in New York. We find Parkinson engrossed in improving a strip
recorder that records voltages on a sheet of paper.

The machine controls a pen held against a strip of paper moving
underneath it. Voltages read by the machine cause the pen to
scratch squiggly lines on the paper; something akin to a lie detector. A small round device, called a potentiometer, controls a set of
magnetic clutches that adjusts the pen. (The round volume control
on your old TV set, for example, is a potentiometer.)

IT CAME TO ME IN A DREAM

Our story is set in the spring of 1940. This date is very important. Everyone was concerned about the rapid drive of the German army through Holland, Belgium, and France. The newspapers were filled with bad news from the war fronts in Europe, especially the news that hundreds of thousands of stranded soldiers had to be rescued from the beaches at Dunkirk, France. This news distressed Parkinson, and the combination of his own engineering problems with problems from the war appears to have been a catalyst for a most vivid and peculiar dream:

> I found myself in a gun pit or revetment with an anti-aircraft gun crew. I don't know how I got there—I was just there. The men were Dutch or Belgian by their uniforms—the helmets were neither German, French nor English. There was a gun there . . . firing occasionally, and the impressive thing was that *every shot brought down an airplane!* After three or four shots one of the men in the crew smiled at me and beckoned me to come closer to the gun. When I drew near he pointed to the exposed end of the left trunnion. Mounted there was the control potentiometer of my level recorder! There was no mistaking it—it was the identical item.[1]

The whole scene faded out. Parkinson woke still retaining a remarkably clear picture of the details. He realized that if his potentiometer could control the high-speed gyrations of a rapidly moving pen with great accuracy, why couldn't a similar device do the same thing for an antiaircraft gun?

THE CRITICAL SKETCH

That morning at work, he talked it over with his boss, Clarence A. Lovell. Lovell was a pioneer in the field of operational amplifiers, analog devices used to add and subtract units of electricity like a digital calculator manipulates numbers. These early computers were crucial to performing the arithmetic calculations needed to gauge the motion of moving airplanes. They immediately set to work. After digesting how the tools he and Lovell had designed could be used to add, subtract, multiply, and divide, Parkinson realized that he'd need to talk Lovell's boss into believing in the dream. So he made a quick sketch of the system on a sheet of typing paper.

Neither Parkinson nor Lovell knew anything about the science of controlling antiaircraft guns; they knew nothing of the crude mechanical gun directors then being used. But Bell Labs executives were convinced. They contacted the Army Signal Corps, who took one look at the idea, fell in love with it, and awarded a contract to the hardware arm of Ma Bell: Western Electric. A working model of the gun director, dubbed the T-10, was delivered to the army for testing on December 1, 1941.

The gun director worked this way. First an optical finder—and later radar—fed altitude, azimuth, and slant range of enemy aircraft to the gun director. The gun director aimed the gun by taking the target's present position and continually recalculating via computer the target's future position. Such heavy computing power in the era of bulky vacuum tubes and wired circuitry took truckloads of men and machinery per gun. But it quickly paid off.

WINNING THE SECOND BATTLE OF BRITAIN

The first production model, labeled M-9, appeared in the field in 1943 and changed the outcome of the so-called Second Battle of Britain. In one month, August 1944, 90 percent of German V-1 buzz bombs destined for London were shot down over the cliffs of Dover. In a single week, radar-guided 90-millimeter guns controlled by M-9s shot down 89 of 91 buzz bombs launched from Antwerp.

Before the end of the war more than three thousand gun directors rolled out of Western Electric factories in Chicago. After the war, Parkinson became a minor celebrity. Accounts of his dream appeared in newspapers and magazines.

The story of the gun director is miraculous enough as is. But scientists and technologists also marvel at another wonder: the positive reception and spirit of cooperation that greeted the gun director dream. Here were two experts virtually unknown outside their chosen fields who dared to bring up a radically new idea based

The M-9 electrical gun director being demonstrated at Bell Labs in 1943. The 90-millimeter gun under control is in foreground. The tracking unit with two operators is shown behind the gun crew. Other units of the system are in the truck at right.

on nothing but an unusual night's sleep. The fact that Parkinson's and Lovell's superiors backed the men's intuition, sought outside help, and were welcomed by experts in the field ranging from mathematics to chemistry is remarkable.

Ordnance experts with no knowledge of the fledgling world of analog computers willingly taught everything they knew to the newcomers. Military and civilian scientists sat down together and coordinated a complex and effective exchange of ideas. That the efforts of a team of people with disparate interests could have been so successful is a minor miracle in itself.

After the war, the technology developed for fire control was used to design a useful new analog computer nicknamed Gypsy. Built from leftover gun control parts, Gypsy operated for over ten years and proved very good at solving problems in science and engineering. So good was Gypsy that it was the basis for the development of the postwar analog computer industry.

22

....

The Typewriter:
The Invention No Office Wanted

*The advent of the first writing machine was not announced in
cable dispatches and newspaper headlines. It slipped into
existence quietly, timidly, unobtrusively, with an indifferent
world to face.*

> —*New York Times,* quoted in *American History of Invention &
> Technology,* Spring/Summer 1988

FOR SOMEONE WHO writes for a living, it's hard to imagine that
anything so necessary to sustain human life as the typewriter would
find itself an orphan when first invented. No one wanted to have
anything to do with it.

The model for all modern typewriters, the original Remington,
was introduced in 1874 to the most deafening silence imaginable.
As the *New York Times* would later document, hardly anyone paid
attention.

But just a few years later the same people who yawned at its
birth were wondering out loud how anyone could have done without
it. It was even being compared to the great inventions of the
century, such as the steam engine. As writer Cynthia Monaco
pointed out, at least one journalist of the day noted with amazement
how "The world needed it. It had always needed it, but it had never
known its need."[1]

The idea of creating a mechanical writing machine was relatively
new. Ever since the advent of movable type the knowledge—the
technology—for building one had already been invented. No great
leaps in science were required. But nobody seemed to have found

The First Secretary. Lillian Sholes, daughter of Christopher Sholes—inventor of the typewriter—shown assuming the pose to be taken by millions of office "girls" of the future.

the need for one until the first part of the eighteenth century. Then in 1714 English inventor Henry Mill patented the first in a long line of writing machines—over a hundred—that would lead up to the first Remington.

Mill never actually built his machine. But his followers proved more adventuresome. William Burt of Detroit patented and built a crude "typographer" in 1829. Being slow and unwieldy, it went nowhere. Along came Charles Thurber in 1843. This Massachusetts inventor patented his "mechanical chirographer," which received some measure of notoriety in *Scientific American*. Thurber mounted his type on a rotating drum, similar to the toy typing machines you can buy in kids' stores. But it worked at such a snail's pace that it proved no match for pen and paper.

Then along came Christopher Latham-Sholes. A Milwaukee inventor whose career included jobs in journalism, printing, and publishing, as well as a tenure as state senator, Sholes never gave up his lifelong tinkering talents. As with many nineteenth-century inventors, Sholes spent his tinkering hours at a local machine shop close to home. He invented two typewriter devices. One printed the addresses of subscribers in the margins of a newspaper to facilitate delivery. The other printed up tickets and coupons. This latter machine was eventually modified to print numbers in the corners of pages in books.

PRINTING FASTER THAN WRITING

Sholes and his colleagues worked at a fertile period in American "tinkerology." Alexander Graham Bell was tinkering with speech, Edison with electricity. And an uncounted few were tinkering with the typewriter. The news of one of those typewriter tinkerers reached the ears of Sholes. A friend of his in the machine shop, Carlos Glidden, showed him an article in an 1867 issue of *Scientific American*, characterizing a machine that can print thoughts twice as fast as they can be written. The machine printed letters by the blow of a minute hammer containing the body of the type, striking a sheet of carbon paper, leaving an imprint on the paper below. Each letter was moved into position in front of the hammer by a key system similar to that of a piano.

Sholes realized immediately that this machine was not much different from his very own page-numbering machine. He improvised a new invention made out of glass, carbon paper, a telegraph key, and lots of piano wire. It could type only one letter—W—and did it so badly as to be impractical. But it was good enough to show investors. Sholes quickly enlisted minimal financial backers, who provided enough money to allow him to improve the machine and patent it in July 1868. But to mass-produce their typewriter, the partners needed to find substantial money.

Sholes turned to an old newspaper friend, James Densmore. After looking at a sample of the machine's type, Densmore fell in love with the idea and bought into the project. He adopted the typewriter as his own idea and prodded Sholes into improving it twenty-five times in the four years following the 1868 patent. In 1870 the first machine was ready for market and manufactured in limited quantity.

THE FIRST REMINGTON

Densmore wanted to have more control of the machine. He pecked away at the other two partners until they sold him their stock in the company. Then it was time to find a serious manufacturer.

He turned to the E. Remington and Sons of Ilion, N.Y. Remington had been an arms producer during the Civil War. With the war over, it had already diversified into the sewing machine business. But the company was still hungry for new products. When Densmore dangled the 1872 model of his typewriter, Remington bit. They agreed to produce thousands of typewriters after reworking the design.

Densmore was elated. He pictured himself sitting in his small office in lower Manhattan and selling his machines to writers and clergymen—literary men. Perhaps small businessmen who needed to send off a letter or two quickly. Never in his wildest dreams did he picture a secretarial pool of hundreds of typists. The typewriter was destined to be used by writers, authors, speech makers: all who enjoyed seeing their words in print. Here was a way to publish one's own printed material without the expense of a full-scale printing press. (Densmore would have loved today's desktop-publishing word processors!)

MARK TWAIN BUYS A TYPEWRITER, BUT NO ONE ELSE

Densmore—the eternal optimist—soon found his idealism put to the test. His first shipment of typewriters, sent by Remington in 1874, were being bought by nobody despite testimonials from some of the country's most famous people. Mark Twain, seeing one in a Boston dealer's window, went in for a demonstration. Shocked by the sight of a woman who could type seventy-five words per minute, Twain decided he must have a typewriter and bought his Remington right on the spot.

But, surprisingly, most people did not find the Remington's typing speed to be a selling point. The machines did not leap off the shelves. Perhaps it was the Remington's price. Carrying a price tag of $125, the Remington No. 1 was beyond the reach of most customers. Perhaps it was the Remington name. Who wanted to buy a fad made by some sewing machine company?

In all of 1874 only a couple of hundred machines were sold, with not much better luck the following year. In 1876 Remington showed off its machine at the Centennial Exposition in Philadelphia. Visitors to the Remington booth were treated to a few typewritten lines they could keep for a mere twenty-five cents (kind of like those circular metal tokens you can stamp your name on at the county fair). As a gimmick, it worked. Everyone queued up to take a look at the newfangled machine. But no one walked home with one.

The same thing happened in New York City. The Remington would draw huge crowds at exhibitions where a typist's flying fingers would spit out words at sixty per minute. Everyone would marvel—oohs and ahs all around—and people would nod their heads and say that typewriters were the wave of the future. But no one would buy one. By the early 1880s only five thousand machines found homes. Densmore and Sholes became paupers, barely able to support their families. The Remington company was in no better shape, constantly flirting with bankruptcy.

No one could figure out why the typewriter didn't catch on. The intangible obstacle lying in its path was the way people perceived the act of communication by letter. Letter writing was viewed as not only the most private form of communication between people but as the most complete and encompassing.

Everything about a letter told you something about the person. The way a letter was addressed, the way it was signed, carried a secret about the writer. Textbooks full of rules dictated the way letters should be written to be socially acceptable. Students attended penmanship school. Quoting from *The Gentlemen's Book of Etiquette*, writer Cynthia Monaco pointed out, "There is no branch of man's education, no quality which will stand him in good stead more frequently than the capability of writing a good letter. . . . In business, in his intercourse with society, in, I may say, almost every circumstance of his life, he will find his pen called into requisition."

Those who dared type risked finding their letters scornfully rejected. A Texas insurance agent received a reply from a customer who complained he did not think it was necessary to have his letters taken to the printer and "set up like a hand bill." Typewritten letters were seen as insulting, implying the recipient was not capable of reading. Since the only kind of mail received set in type was advertising, recipients did not know what to make of typewritten letters. They assumed the letters to be unimportant—akin to junk mail—and threw them away. Imagine arriving at a hotel that had received your typewritten reservation only to find that it had been thrown out with the other junk circulars.

And imagine receiving a letter in which the signature had been typed, too. Hard as it is to believe, even Mark Twain typed his signature on his Remington. How would one know if the letter had been forged? People didn't. That's another reason typing took time to catch on.

Yet the greatest shortsightedness of Remington et al. lay in their inability to see the typewriter as a boon to business. One could see how correspondence was deeply rooted in etiquette and penmanship. But what about bookkeeping? What was so sacred about handwriting numbers? Did a ledger care if it was typed or handwritten?

Densmore and his partners simply didn't target businessmen. Businessmen were not seen as independent agents, guiding a widening sales force that canvassed the country for increased market share (unless you were a railroad tycoon or oil baron). Normally, businessmen were seen as proprietors of small establishments, local shops. And they did not take it upon themselves to embrace the typewriter.

TARGETING BUSINESS

When Remington went after sales, it targeted court reporters, lawyers, clergymen, writers, and editors. Businessmen were an afterthought. In fact, the first dozen typewriters to leave the Remington factory were given away to court reporters in exchange for their endorsement. No one could convince the general public that typewriters would inevitably replace hand correspondence if one ever found its way into an office.

It wasn't until the late 1880s that serious inroads were made. It took the slow depersonalization of the working space following the Civil War to bring on the change. No longer did clerks sit in one spot—like Dickens's Cratchit—and perform all the office tasks of bookkeeping and letterwriting from a cramped candlelit desk top. The telephone, telegraph, and railroads of the latter part of the century turned business to booming. Technology bred prosperity and expanded the work space. The workload was becoming divided. Corporate managers appeared on the scene and began looking for ways to increase productivity.

Recordkeeping became more than any one person could handle. People had to be able to read each other's writing and to quickly fill ledgers with facts and figures. It was in this whirlwind of activity that managers looked down and, lo and behold, rediscovered the typewriter. Here was a device that could turn out fifty words per minute more than a pen. Here was a device that could turn a good typist into a superstar: up to twenty hours of work in one hour. Where had it been all their lives?

THE WORLD'S FIRST TYPISTS

Almost overnight, the sales of typewriters skyrocketed. Where in 1881 Remington tallied a total of twelve hundred typewriters sold, in 1888 fifteen hundred typewriters *per month* were being sold by the new and independent company called Remington Standard Typewriter. The new and improved Remington 2 debuted in 1878, featuring both upper- and lowercase letters.

If imitation is the sincerest form of flattery, Remington was being showered with admiration. By 1885 names like Caligraph, Crandall, Hall, and Hammond graced their own distinctive brands of typewriters with sales of nearly fifty thousand.

In just under ten years, the typewriter went from being a stepchild to the most necessary piece of office equipment. *The Penman's Art Journal* in 1887 noted how the typewriter's "monotonous click can be heard in almost every well-regulated business establishment in the country."

It's often said that success has many fathers; failure is an orphan. In one of the great ironies of history, the success of the typewriter made fathers of people who would just as soon have left the Remington an orphan. Take the comments of one of the early naysayers, a Remington board member. Fifty years after seeing the original Sholes-Glidden typewriter, Henry Harper Benedict erroneously recalled that when first asked his opinion about whether to invest in the new venture, he replied: "The machine is very crude, but there is an idea there that will revolutionize business. . . . We must on no account let it get away."[2]

It's hard to imagine today that people would have no use for such an irreplaceable machine. If you're having trouble believing that yourself, just think about the computer or word processor you have in your office. It won't take another fifty years to wonder how you ever got along without it.

23

The Wasp That Changed the World

*The American wasps form very fine paper . . . They teach us
that paper can be made from the fibers of plants without the use
of rags and linens, and seem to invite us to try whether we
cannot make fine and good paper from the use of certain woods.*

—RENÉ-ANTOINE FERCHAULT DE
RÉAUMUR, November 15, 1719

NAME A UBIQUITOUS product we all take for granted. It's something
almost no one can live—or work—without. Give up? I'm talking
about paper. The stuff this book is made of.

To imagine life without an abundant supply of paper, you don't
have to go back very far. Try the early eighteenth century. Paper
was invented by a Chinese court officer, Ts'ai Lun, in the year 105.
By pounding up plants (probably mulberry bark) with hemp, rags,
and water, Ts'ai created a pulp that he spread over a screen to
drain. After drying the mat in the sun, he found that the resulting
dried sheet could be written on.

THE CHINESE UNLOCK THE SECRET

The Chinese prospered with paper. One emperor's library contained fifty thousand books while Europe remained illiterate. The
Chinese kept papermaking to themselves. But in the New World,
the Mayans, about A.D. 500, stripped bark from fig trees, beat it to
soften it, treated it with lime to remove the sap, and made writing
materials. The Aztecs improved this method and found paper so

valuable they used it as tribute. Only when the Moors of North Africa captured Chinese papermakers in 751 did papermaking spread to Europe. Though the Chinese had invented paper hundreds of years before, Europeans couldn't think of what to do with it. There weren't very many books. Those that did exist were printed on papyrus leaves, parchment, or paper made from rags.

Then along came Johannes Gutenberg and his movable type invented in the middle of the fifteenth century. Certainly that created a tremendous demand for paper. But paper in those days was made from cloth—old rags—and while looms were set to weave as much cloth as they could, just so many shredded garments could be set free to meet the ever-increasing demand for paper.

Even until the early nineteenth century, making paper was a slow and tedious process. Each sheet of paper was made by hand, using a technique not much improved from the ancient Chinese method practiced almost two thousand years before. A good paper-maker could make only about 750 sheets per day.

The turn of the eighteenth century saw demand for rags skyrocketing as people found many more uses for paper.* The invention of mechanical pulpers or beaters outstripped the supply of rag fibers. Papermakers used any rags they could get their hands on, and that meant that the quality of paper suffered from one batch to another.

AMMUNITION FROM BOOKS

When American Revolutionary War soldiers ran out of paper wadding for their barrel-loaded guns, they ripped up old books. John Adams complained to his wife in a letter: "I send you now

---

* Paper was becoming popular as a building material. An Englishman, Henry Clay, applied for a patent in 1772 to use laminated paper in making panels for floors, walls, cabinets, wheeled coaches, tables, chairs—you name it. And in 1793 a small town in Norway even built a church out of paper!

and then a few sheets of paper; this article is as scarce here as it is with you."[1]

In 1789 technology, seemingly coming to the rescue, just made matters worse. For in that year paper production became mechanized. Nicholas Louis Robert, a clerk at a paper mill in Essenay, France, drew up plans for a machine that would make paper in continuous rolls. The soul of this new machine was an endless loop of wire screening that, when cranked by hand, filtered the pulp and made one long sheet of paper.

Unable to come up with the capital needed to see the project through, Robert sold the patent to the Fourdrinier brothers in London.

A FORTUNATE WALK IN THE WOODS

Things looked bleak for the paper industry until the day a famous French scientist took a walk in the woods and along the way solved the paper crisis.

Paper Pioneer. René-Antoine Ferchault de Réaumur, eighteenth-century French scientist, wondered how wasps could make paper (nests) from wood when people couldn't. While there is no record that he ever made paper from wood, he was the first scientist to suggest it.

René-Antoine Ferchault de Réaumur was one of the greatest scientists France has ever produced. Born in La Rochelle in 1683, Réaumur developed into a world-class physicist, mathematician, and chemical engineer. He mastered all three disciplines in one year at the age of twenty. His knowledge was widely sought out in Europe, and the world's greatest scientific societies, the Royal Society of Great Britain, the Academies of Sciences of France, Russia, Prussia, and Sweden, and the Institute of Bologna all welcomed him as a member. His investigations (1722) of the production and treatment of steel proved invaluable to France's antiquated ferrous metal industry. One of his most famous accomplishments was the invention of a thermometer that later bore his name.

Lucky for us, Réaumur found time among his real work to pursue his favorite hobby—insects. He enjoyed entomology more than anything else and would spend hours going through the innards of birds looking for traces of their insect diet. Amid his experiments

Paper Wasp's Nest. The wasp uses dry wood—shingles, boards, fence rails—which it scrapes off and chews up. Mixing the wood with its natural body fluids, it forms a paste that is spread out to dry in layers. The nest remains dry due to the natural "shingling" effect.

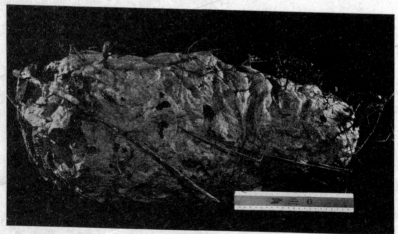

on birds and insects, Réaumur published in Amsterdam a six-volume work entitled *Mémoires pour Servir à l'Histoire des Insectes*, from 1737 to 1748.

Take this brilliant scientist and accomplished author and place him smack dab in the center of our paper crisis. As an author he is acutely aware of the problem of getting his works printed. As a chemist he knows everything there is to know about making paper. And as a naturalist he does the most natural thing to clear his mind: He takes a walk in the woods. Here we have Pasteur's classic ingredients for discovery: the prepared mind meeting the fortuitous event.

HOW WASPS DO IT

We'll never know the exact date that Réaumur took his famous walk. But that's not important. What is important is what this renaissance man happened to observe. As he walked through the woods, an abandoned wasp's nest caught his eye. Here was a discovery no entomologist could pass up; a chance to examine the lair of a stinging insect when no one was home. Peering closely at the honeycombed structure, Réaumur must have been dumbstruck. Something in his brain went *click*. At that moment he realized that the wasp's home was made from *paper*. Crude paper for sure, but paper strong and sturdy enough to withstand exposure to the elements.

And the wasp made the paper without rags! How could this be done? What did the lowly wasp use for simple raw materials that had so eluded the human brain? The answer lay at his feet and stared him in the face. Wood. The wasp built his paper from the small twigs found in the forest. But without chemicals and fire and mixing tanks, what did the wasp use to fashion its paper? Where was the factory? The answer was elementary. The wasp processed the wood into paper in the only crucible it had: its stomach.

What a discovery! How many entomologists had taken the same walk in the woods and observed a wasp's nest? Countless numbers. But how many of them knew enough about the digestive systems of animals to take advantage of the discovery? Here was something Réaumur was uniquely qualified to study. Réaumur was already the world expert on the digestive system of birds. Why not take a look at the digestive juices of the wasp?

PAPER FROM WOOD

Over the next few months, Réaumur sought a way of explaining how the wasp bit off tiny pieces of wood from trees, swallowed it, digested it, and excreted it in the form of paper. He accomplished the feat in grandiose manner, learning just about all there was to know about the industrious insect. Publishing his findings on November 15, 1719, Réaumur shared with the French Royal Academy what he had learned about the wasp:

> The American wasps form very fine paper, like ours; they extract fibers of common wood of the countries where they live. They teach us that paper can be made from the fibers of plants without the use of rags and linens, and seem to invite us to try whether we cannot make fine and good paper from the use of certain woods. If we had woods similar to those used by the American wasps for their paper, we could make the whitest paper, for this material is very white. By a further beating and breaking of the fibers that the wasps make and by using the thin paste that comes from them, a very fine paper may be composed. . . . This study should not be neglected, for it is, I daresay, important.[2]

Réaumur marveled at the wasps' workmanship, especially the paper made by the "species of wasp that lives in Canada." He was so impressed with the wasps' craftsmanship that he remarked:

Even after examining the surface for a considerable amount of time, one would accept the next as the work of the hand of man. Its covering resembles our paper to such an extent that it is hard to detect a difference. It has the same gloss and the colour is that of an old piece of manufactured paper which has formerly been white. It is fine paper and as heavy as that of ordinary portfolios.

Despite his observations, if Réaumur ever duplicated the wasp's feat he left no record of it. He lamented this fact later in life, 1742:

> I am ashamed that I have not yet tried this experiment, it is more than twenty years since I first realized its importance and made an announcement of it. But I had hoped that someone would have been interested in making it his occupation.

Réaumur died without ever having made a sheet of paper from wood. But his plea set Europe looking for new sources of raw material for paper. Albert Seba, a Flemish natural historian, compiled a set of books in 1734–65 in which he suggested that "alga marina"—seaweed—be used to make paper. Jean Étienne Guettard of the Royal Academy of Sciences proposed in 1741 that paper be made from swamp moss. In 1764 English diplomat John Strange suggested the use of broom (a shrub) as a papermaking material.

POTATO PAPER?

Much of this was all talk and no action. But then a clergyman from Germany read of Réaumur's work and decided to put it to the test. Jacob Christian Schäffer gathered eighty different kinds of vegetables and tried making paper from them. First he chopped the vegetables by hand and then later invented a machine to grind them up. He published his works in a six-volume set between 1765 and 1771. Most of the experiments were carried out in his home. And the results can be found bound up in his works; specimens of paper made from wasps' nests, straw, cabbage, asbestos, pine cones, potatoes, and old shingles, to name a few.

THE FIRST WOOD TRIALS

The first person to actually try to make wood-based paper commercially was Mattias Koops, a Hollander living in London. In 1850 he set up the first paper mill that processed a combination of wood and straw to make paper. He received two British patents for papermaking using a variety of natural woody and vegetable ingredients. And while he successfully produced paper from wood, he couldn't sell any of it. The conservative public just wouldn't buy anything not made from linen and cotton. Koops's business went belly up.

Europe took decades to adjust. A breakthrough came in 1840. Friedrich Gottlob Keller, a German weaver, read Réaumur's work and secured a German patent for a wood-grinding machine. His device pulped wood by grinding it away with a spinning grinding stone. But the paper it made was of inferior quality compared to rag pulp. Nevertheless, the idea was a good one and needed to be developed. Keller's patent was bought up in 1846 by another German, Heinrich Voelter, who built production models of the machine. In 1866 the first American version of grinding machines appeared in the United States. Based on the Keller-Voelter patent, this wood-grinding mill was established in Curtisville, Mass. (now Interlaken) near Stockbridge in the Berkshires.

The first wood-pulp newspaper in the United States was the January 7, 1868, edition of the *Staats-Zeitung*, a German-language newspaper in New York. The *New York World* and the *New York Times* followed suit. But it wasn't until the 1880s that wood-pulp paper was more widely used than all-rag papers.

Today more than 90 percent of our paper is made from wood. We owe it all to a fortuitous set of circumstances: the right person in the right place at the right time some 250 years ago.[3]

24

....

The First Video Game: If You Build It They Will Come

THE WORLD'S PLAYING fields have served as more than the sites of great sporting events. Some of the most memorable milestones in the history of this century have occurred in sports arenas when sporting events were absent. On the road to the atom bomb, scientists built the world's first nuclear reactor at a football stadium at the University of Chicago, under the stands at Stagg Field. It went on line December 2, 1942.

Every student of history knows this story. But how many are aware of the historic event that occurred at another playing field, this one located at one of the country's leading nuclear labs? Unlike the stands at Stagg Field, this site witnessed a historic event that has gone virtually unnoticed. No plaque commemorating the event has ever been put up; no encyclopedia has included the place—or the event—in its hallowed text.

Yet every day at noon dozens of lunchbreak athletes gather to play basketball on a court that served as the site of an important leap forward in modern technology, an event that would influence our lives for decades to come. When asked between dribbles if they know of the historic nature of the playing site, most athletes try to name a historic game. Was it Harvard vs. Yale? Could it be the Celtics against the Lakers? No. No famous athletic competition occurred beneath these hoops more than thirty years ago. Few if any can correctly answer, "This is where the world's first video game was played in 1958."

Decades before Pac-Man swallowed his first dot, eons before Space Invaders were blasted from video screens, and generations

213

before the Mario Brothers opened shop, the world's first video game was quietly born. Without fanfare, without hype—even without a marketing plan—the video game slipped quietly into existence as a last-minute entry in the history of the quiet little Long Island town of Upton.

These days video games are introduced with multimillion dollar promotions: tie-ins to Saturday morning cartoons and free offers on overly sweet kids' cereals. (Of course, you finance this schlock by agreeing to pay the exorbitant prices for glorified designer chips of sand.)

But back in the good old innocent days of the 1950s, when the most exciting addition to TV twin-leads was a rotary antenna, America existed, even thrived, without video games. No Nintendo. No Ninja Turtles.

All that changed one day in 1958 when physicist Willy Higinbotham, from Brookhaven National Laboratory in Upton, got bored with what he was doing and invented the world's first video game.

BORN IN THE NUCLEAR AGE

Remember that name Higinbotham, because years from now Higinbotham will become as famous as the other legendary names in sports. What Naismith was to basketball and Doubleday to baseball (rightly so or not), Higinbotham will be to video games. (I'm doing everything I can, Willy, to get you that recognition you deserve!)

Higinbotham was no newcomer to invention. Willy, as all his friends call him, was a very serious scientist. (A physicist on the Manhattan Project, he observed the explosion of the first atomic bomb.) The whole affair started innocently enough. Each year in the late fifties, Brookhaven Lab would hold an open house. In the post-*Sputnik* era, America was in the grip of an intense nuclear-bomb mania. The government had printed up booklets describing how to survive a nuclear attack in your own hand-built fallout shelter. Being a government nuclear research facility devoted to the peaceful use of atomic energy, Brookhaven Lab decided it could best stem the nuclear hysteria of the day by inviting the public into the labs to see that its employees didn't actually glow in the dark.

Parents and kids filed by the traditional radiation detectors, complicated electronic circuits, and newly hatched brine shrimp (always good to have an animal on display). Rather boring black and white pictures showed peaceful research under way at the lab. Not much happening among the cardboard and wood stand-up displays. How many times can you watch a shrimp wriggle under a light bulb?

Willy, who headed the lab's instrumentation division, was as bored as the tourists. And in 1958, Willy couldn't stand it anymore. Stretching his full five-foot, four-inch frame, Willy decided he'd concoct something more interesting. Something to bring people in on Saturday. Something that moved.

BUILT FROM SPARE PARTS

Putting his Cornell physics training and his MIT electronics background to work, Willy hit upon the idea of building an exhibit around that new and wonderful mesmerizer of the public, television. Only this time Higinbotham didn't need the whole TV set, just the part that people watched: the tube. In fact, he didn't even need to use a real TV picture tube, but the TV's ancestor: a cathode-ray tube from a laboratory test instrument called an oscilloscope.* Dusting off an instruction manual packed with the lab's small analog computer, Willy found directions for hooking the scope to the computer and making a bouncing ball appear on the screen. Would it be possible to control the action of the ball at will? Willy thought so. Scrounging around the lab for assorted resistors, capacitors, and potentiometers, Willy wired the first crude video game out of spare parts.

"We looked around and found that we had a few pieces we could throw together and make a game," recalled Willy years later,

---

* As a test instrument, the oscilloscope is a way to view complicated TV signals or simple AC house current. If you were a keen observer of science fiction on TV in the 1950s and 1960s, you've seen an oscilloscope in the background of a laboratory. It has lots of squiggly sign waves running across its screen; it signifies laboratory "reality." "The Outer Limits" highlighted an oscilloscope in its opening scene.

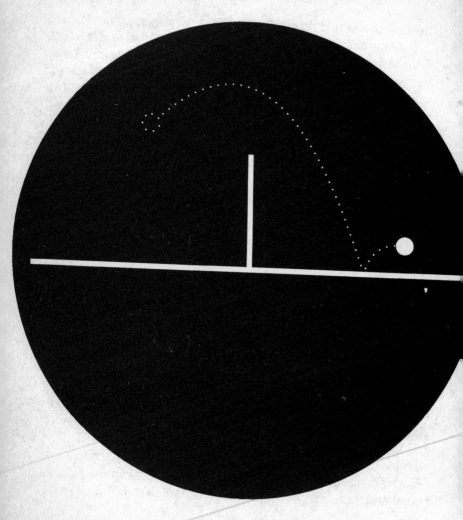

Tennis, Anyone? Here, shown actual size, is a reproduction of the screen of the world's first video game. A tennis court, shown looking from the side, serves as the playing field. A white ball is hit back and forth over the net. Each player has a box with a button and a knob. The button serves the ball and the knob controls the ball's angle. A ball hit into the net would actually bounce off—an unheard-of video effect in 1958. Unlike Pong, which came later, this game showed no paddles on the screen.

"which would have a ball bouncing back and forth, sort of like a tennis game viewed from the side."

Two hours of scratching a design on paper and two weeks of wiring and debugging, and the game was complete. The tennis game was displayed on a tiny, five-inch screen. It involved two players, each having a box with a button and a knob. If you pushed the button, you hit the ball to the opponent's court. The knob controlled how high the ball was hit.

"You could actually hit the ball into the net," recalled Dave Potter, Willy's associate, "see it bounce into the net and see it bounce onto the floor back to you. So it was a very cleverly designed game Willy came up with."

AN UNBELIEVABLE SUCCESS

Set up on a table in the gym right under the basketball hoop on the far wall, the tennis game was practically lost among the other

World's First Video Game. No, not the entire setup. Just the small round screen near the left of the picture and the two boxes below it. The game sits cramped among electronic test equipment and an early digital computer, Merlin, at the Brookhaven National Laboratory. Hundreds of visitors waited in line to play this crude tennis video game in 1958.

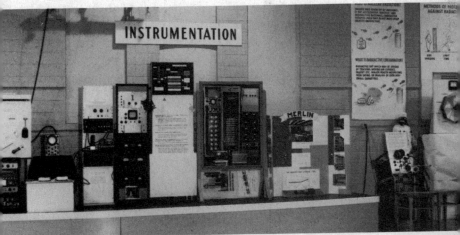

electronic gadgets on display. But when the doors to the gym were opened, the public had no trouble finding the video game. Willy and Dave could hardly believe their eyes.

"Willy's tennis game was a hit," remembered an astonished Potter, "there were long lines. People wanted to play."

Willy was shocked. What was so attractive about a dot bouncing around an oscilloscope screen?

"It never occurred to me that I was doing anything very exciting. The long line of people I thought was not because this was so great but because all the rest of the things were so dull."

For the following year's (1959) open house, Willy improved the game. The picture tube was enlarged to ten or fifteen inches and a novel feature was added: Visitors could play tennis on the moon with very low gravity or on Jupiter with very intense gravity. Again the game was a great success; hundreds stood in line to play.

Most fathers of such a success might have immediately set out to cash in on their invention. But not Willy. Despite the game's popularity, Willy never made a nickel off his invention. He never saw any commercial value in it. The idea seemed too obvious. Anybody with simple equipment bought at Radio Shack could make the game. So Willy never patented the idea.

Years later, Willy told me, "My kids complained about this and I keep saying: Kids, no matter what, even if I patented it, I wouldn't have made any money." Because he worked at a government institution, Uncle Sam would have owned the patent and Willy might have collected a ten-dollar royalty. And anybody wanting to market a video game would have had to pay the U.S. Treasury royalties.*

"I never realized it until many years later," Willy said in 1983, "when the first dumb games came out about 1970–71, that these [video games] would be as popular as they have turned out to have been."

---

* Can you imagine the dent video game royalties would have made in the national debt if every Nintendo, NEC, and Atari sale put a few cents into the federal coffers?

KING PONG

About that time, 1971, video games were beginning to make their first commercial appearances. Pong, a video arcade game marketed by Atari, Inc., was becoming the hit of the bar scene. In Pong, a ball was hit back and forth over a net by two players. (Looking amazingly like Willy's tennis game, Pong would become known as the granddaddy of video games.) Odyssey, the first home video game, was also being released by Magnavox.

According to Higinbotham, the patent for the first video game was applied for by Sanders Associates in 1964. It was purchased by Magnavox, who realized there was real money to be made here. Of course where there's money, there are lawyers. Armed with the patent, Magnavox's lawyers set out to sue all other video game competitors. It was just a matter of time before the legal mess found its way to Willy's door.

"About 1976," wrote Willy in an article for the quarterly *Alumni Review* of Williams College in 1984, "a patent lawyer for a competitor found someone who had played the game at Brookhaven and tracked me down. By good luck, we had made drawings of the system. Thus began an on and off relationship with patent attorneys."

Lawyers have been interviewing Willy ever since. Over the last fifteen years, they've cross-examined him, had him sign affidavits, telephoned him to confirm or deny rumors. In 1981 one of Magnavox's competitors (Willy's memory is a bit foggy on this) figured it had enough ammunition to break the patent. A trial was scheduled for June 1982 in Chicago. At the last moment the case was settled out of court. Magnavox made the competitor an offer it couldn't refuse. The competitors figured they'd better settle.

Disappointed, Higinbotham lost his chance to put his story before the public. "We didn't get our day in court," he wrote. But the legal battle was a learning experience. After watching the enormous legal fees change hands, Willy learned how silly he had been for not patenting his video game. A holder of twenty patents, Willy realized that holding one more couldn't hurt.

"At my deposition, one of the lawyers said that he was very sorry to see the end of an enterprise that had paid him well for five years. It was then obvious that I should have applied for a patent." Willy finally did testify in court in San Francisco in 1985 in a case pitting Activision against Magnavox.

Willy is getting his share of public (vs. legal) recognition now. Too much, he says. His video-pioneering role came to light in 1982 in an article written in *Creative Computing* magazine. That, and a story I did for National Public Radio's "All Things Considered" in 1983, have brought him instant fame.

"I get letters from friends I haven't seen in thirty years who say 'Hey, Willy! I didn't know you did that,' or 'I remember [when it happened].' So that's kind of nice. But it's not the only thing of any means that I've done in my life."

Wouldn't it be ironic, I suggested to Willy in an interview, if fifty years from now when historians write about the history of Brookhaven National Laboratory, they overlook all the famous high-tech radiation work that has gone on there and point to the spot on the gym floor where the video game stood?

"Nobel Prize winners are going to be awfully disappointed if that happens," he said jokingly.

OUT OF RADAR CAME VIDEO GAMES

If Willy is the father of the first video game—and no one else is staking that claim—it's because, like Bell with his telephone, Willy was properly prepared, in the Pasteurian sense of the word, to invent it.*

In January 1941 Willy joined the staff of the MIT Radiation Laboratory. He was assigned to help develop radar, specifically the cathode-ray tube displays. "I worked on circuits to display the radar data on cathode-ray tubes. I authored several patents for the

---

* Louis Pasteur, when asked what role luck plays in discovery, said: "Chance favors the prepared mind."

government." His picture tube work did not differ substantially from the problems involved in the tennis game display.

Before going to Los Alamos, New Mexico, in December 1943 to work on development of the A-bomb, Higinbotham worked on advanced radar displays for the high-flying B-28. To perfect this radar, Willy invented and obtained patents for special kinds of amplifiers like those that later showed up in the analog computer that he used in the tennis game.

So it was not by accident that Higinbotham envisioned how the tennis game would work. Just the opposite. Willy had spent the last twenty years preparing his mind to make the leap. He had invented key elements of it years before. Key pieces of his invention had been coming together for over two decades. The puzzle assembled itself in Willy's mind in 1958. To Willy, inventing the video game was like reinventing the wheel: "It was a natural progression for me."

REMEMBERING THE FIRST VIDEO GAME

No plaque has been erected in the Brookhaven gym where the first video game was played. You won't find the game stored in the attic of the Smithsonian; Willy dismantled the gadgetry after the exhibit ran for two years.

If Willy has his way, he hopes he's remembered for more than the video game. What leaves the sharpest image in Willy's mind is the day he witnessed the detonation of that first atomic bomb at Los Alamos. As designer of the timing device for the bomb, Willy watched the explosion from his vantage point twenty-four miles from ground zero. Since that blast Willy has devoted his energies to making sure another one never goes off. As senior scientist at Brookhaven, Willy spent years amassing the world's biggest library devoted to nuclear safeguards. He has spent the last forty-plus years working on arms control. It's his arms control achievements that Willy hopes he will be remembered for at Brookhaven.

Epilogue: Who's Laughing Now?

Does your neighbor stay up late, night after night, but with no signs of partying? Does he or she remain closeted in the basement, attic, or garage oblivious to the Super Bowl? Do wild sounds of joy followed by angry howls of frustration get your attention? You may be living next to an amateur inventor.

Scattered in basements, garages, and laboratories around the world today's inventors toil in their spare time to fashion their dreams into reality. Your neighbor might be one of them.

Inventing is a very tough business. It takes a thick skin and the ability to keep smiling after constantly hearing the word no. An inventor's days are filled with polite praise about the wonderfulness of one's invention but the sad reality that one's prized creation is "just not a good fit with our company . . . at the moment."

The truth is that very few giant American companies are actually interested in genuine innovation. Yes, thousands of salaried inventors work on the staffs of corporate laboratories. And many churn out small patentable inventions for which they get no royalties. But when an inventor comes up with a *really* new solution to an old problem or a better way of producing an item, few companies welcome the idea.

Just ask the inventors who try to peddle their patents. They can't talk companies into investing in them. Many are forced to take their ingenuity abroad, where new ideas are more welcome. Jacob Rabinow, an inventor with over two hundred and twenty American patents, and a hundred foreign ones, told me that some of his most successful ideas could not be marketed in this country because the firms here found them too "different"—meaning untried by a competitor and therefore risky.

Big companies, in general, would rather sit back and watch another smaller company try its hand at something, and then if successful buy out that company or corner the market with its own brand. (Think of the story of Apple Computer. IBM was not even a player in the multibillion–dollar microcomputer field until years after Apple's two fledgling founders designed their first computers in a garage.)

Why such an aversion to innovation? Rabinow says that years ago, American companies were smaller and owned and run by an inventor/entrepreneur who usually stayed on as chief executive. Such an entrepreneur devoted his life to running the company. America's biggest and best research labs were founded and funded by men like David Sarnoff at RCA or Thomas J. Watson and his son Thomas Junior at IBM, who spent sixty years between them running the company. These men knew their company top to bottom and bet their reputations on it. Quality became paramount.

"Today most companies are run by financial executives," laments Rabinow, "who know how to buy and sell each other. But they don't know anything about the products they make. They may not even know *what* products they make."

With the death or departure of the founder, quality often gives way to quantity. Form follows finances instead of function. Products become commodities, no different in quality and design than any other. Research and development takes a back seat to marketing.

Corporate executives come on who have no personal interest in what they make. New ideas take many years to yield results. From drawing board to prototype to product can take three, four, five years or more. But the corporate lifetime of the financial executive in charge of the project may be much shorter. That executive may be more concerned with showing a profit for the next quarter, or may have moved on to another company and not be around to see it bear fruit.

Or a company may be just too big and fat to change the way it does business. Corporate management may decide there is no point in doing what other companies are doing. Two examples. RCA invented color television and wanted to sell color TV sets. But it couldn't get any of the television stations interested in broadcasting color television signals. They just laughed. What's the advantage of watching "I Love Lucy" or any of the 1950s and early 1960s TV shows in color. Who needs it? And besides, they argued, since RCA is the only game in town manufacturing color TV equipment and owning the patents, we'd be loading the company coffers.

What to do? David Sarnoff single-handedly forced broadcasters to adopt color television by cleverly flooding the market with color sets. He sold RCA sets to the public for less than what it cost to make them: four hundred dollars for a seven-hundred-dollar set. He then encouraged people to call or write their local TV stations and ask them why they weren't supplying color programming to their new (read: expensive) sets. Meanwhile, the NBC-TV network, which was owned by RCA, began color programming and adopted a colorful peacock as its symbol. An announcer proudly crowed before a display of animated, unfolding feathers that "the following program is brought to you in *living color* on NBC." Color TV sales zoomed. And no one laughed at David Sarnoff.

NOT INVENTED HERE

Why are big companies the last to gamble on innovations? Sometimes their management is paralyzed in its plush seats. Sometimes they fall victim to the "not invented here" syndrome. That's what happened to the Ford Motor Company. Big companies like Ford and GM take years to make decisions; they get locked into doing things their own way and find it hard to change. So when a new idea is brought to them, they look right down the gift horse's throat. For example, Henry Ford II related how immediately following World War II the British had invited him to come to Germany. As a war reparations payment the British offered Ford a potential gold mine: the entire Volkswagen automobile plant and all related patents. Ford turned them down. He was not interested in making a "junky" car like that, rotten in design. Twenty-five years later, when Volkswagen was making a huge dent in the American car market, Henry Ford II told the German magazine *Der Spiegel* that if he had to do it all over again, he would still make the same decision. He would still refuse the offer. He wasn't interested in any kind of change from the vision Ford had of what a car should be. Despite the demonstrated popularity of the Bug, despite being what consumers wanted, the Volkswagen idea was *not invented here*.

If Japanese corporations had become infected with "not invented here" disease, we'd all be watching American-made VCRs playing on American-made TV sets. Take the transistor radio. The transistor is an American invention, invented at Bell Laboratories (AT&T) in 1947. In accordance with Bell Labs policies of keeping its research open to the public, Bell Labs began granting licenses to anybody who wanted to produce and market transistors. In 1952, to disclose the technology to those companies, it sponsored a week-long seminar on how the transistor worked. Twenty-six domestic and fourteen private foreign licensees attended at a cost: a twenty-five-thousand-dollar license fee to get in. A steep price for casual browsers, but a bargain for serious companies who could credit the admission fee as an advance against royalties to the Bell System. Texas Instruments, a small electronics firm that made military and geological equipment, mailed a check immediately. Teaming up with another tiny company, the Regency Division of IDEA, Texas Instruments contracted with Regency to produce the first transistorized pocket radio, the TR-1. Rushed into completion for Christmas 1954, the world's first pocket transistor radio, the Regency TR-1, became an immediate hit. About a hundred thousand were sold the first year.

The licensing fee was very steep for a tiny Japanese company, the Tokyo Telecommunications Engineering Corporation and its bright co-founder, Akio Morita. Morita, a brilliant young physicist, had started his firm in a bombed-out department store in Tokyo. His partner, Masaru Ibuka, learned from a friend in the United States that the Bell System was licensing its transistor patent. No sooner had the small company scraped together the license fee than it was met by a wall of Japanese government red tape. The government, overly concerned with letting so much money leave the country, refused to approve the license. So Morita personally interceded. In 1953 he paid a visit to the United States, convinced the Bell System to grant him a provisional agreement, and rounded up every book he could on transistor technology. His industriousness impressed the Japanese government. In January 1954 it allowed the license to go through. Much to the surprise of the government in Tokyo, no other Japanese company had shown the least bit of interest in the transistor.

Ibuka and a hand-picked team of physicists flew immediately to New Jersey and soaked up as much knowledge as they could at Bell Labs. Each night they mailed back to Japan everything they had learned. The hard work paid off.

By June 1954 the first Japanese transistor was born at Tokyo Telecommunications. News of the success of the Regency TR-1 sped up the development of the transistor radio. August 1955 saw the birth of the first Japanese transistor radio. It was called Sony, which in 1957 became the new name of Tokyo Telecommunications.

By the late 1950s Sony's competitors, Toshiba, Hitachi, and Matsushita, were all marketing pocket transistor radios. And the world was soon flooded with Japanese radios. "Made in Japan" became synonymous with miniaturized transistor radios.

Imagine if Sony had shunned the transistor as an "American invention." It's no wonder that foreign companies, like those in Japan, are threatening to dominate technology: Their corporate executives are willing to take chances, to innovate, to learn from others. And they've learned their lessons by studying the history of their big, American brothers, thick-skinned inventors like Bell and Edison.

There is one ironic note. Turning down innovative ideas is not a new practice but an old and timeless story. Only a hundred years have passed since Western Union laughed at Bell when he offered them the telephone. Who's laughing now?

Source Notes and Further Reading

CHAPTER 1

1. Bern Dibner, *Benjamin Franklin, Electrician*, Burndy Library, Electra Square, Norwalk, CT, 1976, p. 19.
2. Ibid., p. 22.
3. Ibid., p. 24.
4. Ibid.
5. Ibid., p. 25.

CHAPTER 2

1. Any look at Edison would not be complete without the following outstanding books: Arthur A. Bright, *The Electric Lamp Industry*, (New York: 1949); Robert Friedel and Paul B. Israel, *Edison's Electric Light: Biography of an Invention* (Brunswick, N.J.: Rutgers University Press, 1986); Robert Conot, *A Streak of Luck* (New York: Seaview Books, 1979); and Mathew Josephson, *Edison* (New York: McGraw-Hill, 1959).

   In addition, the following article is wonderful reading: George Wise, "Swan's Way: Inventive Style and the Emergence of the Incandescent Lamp," *IEEE Spectrum*, New York, 1982.

CHAPTER 3

1. Mathew Josephson, *Edison* (New York: McGraw-Hill, 1959), p. 345.
2. Letter to committee member Dr. Alfred P. Southwick, December 1887, Edison Archives, Edison National Historic Site, West Orange, New Jersey.
3. Terry S. Reynolds and Theodore Bernstein, "Edison and 'The Chair,'" *IEEE Technology and Society Magazine*, March 1989, p. 21. Reprinted from *Electrocution Hearing*, 1: 396–97.
4. Ibid., as reprinted from *Electrocution Hearing*, 2: 630–31.
5. Ibid., p. 21.

CHAPTER 5

1. Curt Wohleber, "The Bandleader's Blender," *American Heritage of Invention & Technology* (New York: American Heritage, Fall 1989), p. 64.
2. Ibid.

CHAPTER 8

1. Robert V. Bruce, *Bell: Alexander Graham Bell and the Conquest of Solitude* (Boston: Little, Brown and Company, 1973), p. 36.
2. Ibid., p. 182.

CHAPTER 9

1. Donald G. Fink, "The Tube," *Science 84*, American Association for the Advancement of Science, Washington, D.C., November 1984, p. 123.
2. Erik Barnouw, *Tube of Plenty: The Evolution of American Television* (New York: Oxford University Press, 1990).

CHAPTER 11

1. For further reading, consult Dean J. Golembeski, "Struggling to Become an Inventor," *American Heritage of Invention & Technology* (New York: American Heritage, Winter 1989), pp. 8–15; William Copulsky, *Chemtech* Magazine, American Chemical Society, Washington, D.C., May 1989, p. 280; and Xerox Corporation's *1990 Fact Book* (Stamford, CT: Xerox Corporation), p. 1.

CHAPTER 12

1. Charles H. Townes, "Harnessing Light," *Science 84*, American Association for the Advancement of Science, Washington, D.C., November 1984, p. 153.
2. Ibid., p. 155.

CHAPTER 15

1. For further reading, consult David A. Hounshell and John Kenly Smith, Jr., "The Nylon Drama," *American Heritage of Invention & Technology* (New York: American Heritage, Fall 1988); and the chapter "Nylon: Cold Drawing Does the Trick," in Royston M. Roberts's *Serendipity* (New York: John Wiley & Sons, 1989).

CHAPTER 17

1. Bernard E. Schaar, "Chance Favors the Prepared Mind," *Chemistry* Magazine, American Chemical Society, Washington, D.C., March 1968.

CHAPTER 18

1. Earl L. Warrick, *Forty Years of Firsts* (New York: McGraw-Hall, 1990), p. 28.

CHAPTER 19

1. John F. Shoch and Jon A. Hupp, "The 'Worm' Programs—Early Experience with a Distributed Computation," *Communications of the ACM*, March 1982, 25: 172–180.

CHAPTER 20

1. From an unpublished manuscript by Richard Q. Hofacker, Jr., quoting *The Electrical Engineer* (New York, May 12, 1898).

CHAPTER 21

1. W. H. C. Higgins, B. D. Holbrook, and J. W. Emling, *A History of Engineering and Science in the Bell System: National Service in War and Peace* (Murray Hill, N.J.: Bell Telephone Laboratories, Inc., 1978), p. 135.

CHAPTER 22

1. Cynthia Monaco, "The Difficult Birth of the Typewriter," *American Heritage of Invention & Technology* (New York: American Heritage, Spring/Summer 1988), p. 12.
2. Ibid., p. 21.

CHAPTER 23

1. Quoted in the American Paper Institute's pamphlet *Paper and Paper Manufacture* (New York: 1987), p. 2.
2. Quoted in *Great Accidents in Science That Changed the World* by Jerome S. Meyer (New York: Arco Publishing Company, 1967), p. 50.
3. For further reading, consult Dard Hunter, *Paper Making Through Eighteen Centuries* (New York: Burt Franklin, 1930).

Index

Index

Index

The author gratefully thanks the following for permission to use their photographs and illustrations:

American Institute of Physics, Niels Bohr Library: page 178
American Museum of Natural History, Department of Library Services (H. S. Rice, photographer): page 208
AT&T Archives: page 87
AT&T Bell Laboratories: pages 79, 195; also pages 3, 4, and 7 of second photo insert
Binney & Smith Inc.: page 166
Brookhaven National Laboratory: page 217
Burndy Library: pages 7, 33, 64, 207; also pages 1, 3, and 4 of first photo insert
Computer Museum: page 171
Du Pont Corporation: pages 136, 137, 139
Eastman Kodak Company: pages 48, 50; also pages 5 and 6 of first photo insert
Harry K. Flemming (courtesy Burndy Library): page 2 of first photo insert
Hagley Museum and Library: pages 150, 198; also pages 5 and 8 of second photo insert
Hughes Aircraft Company: pages 123, 124, 125
ITU Telecommunication Journal: page 63
Dan Keller: page 66; also page 8 of first photo insert
New York Public Library, Astor, Lenox and Tilden Foundations, Rare Books and Manuscripts Division, Chester Carlson Papers: page 114
Paterson Museum: page 190; also page 6 of second photo insert
Pennsylvania State University, Fred Waring's America: page 55; also page 7 of first photo insert
Raytheon Company: pages 59, 61
Smithsonian Institution: pages 72, 95; also page 1 of second photo insert
University of Pennsylvania: page 169
University of Wisconsin: pages 38, 41
U.S. Department of the Interior, National Park Service, Edison National Historic Site: pages 12, 19, 22

U.S. Navy, Naval Surface Weapons Center, Dahlgren, VA: page 180
VELCRO® U.S.A. Inc.: pages 132, 133
David Wrubel: page 89

The illustrative material on pages 105, 107, and 108, as well as on page 2
of the second photo insert, is taken from *Baird of Television* by Ronald
Tiltman (Seeley Service & Co., Ltd.: London, 1933).
Drawings on pages 118 and 216 are by the author.